THE BIOLOGY AND ADAPTABILITY
OF NATURAL POPULATIONS

JAMES T. GIESEL

Department of Zoology, University of Florida,
Gainesville, Florida

THE BIOLOGY AND ADAPTABILITY OF NATURAL POPULATIONS

with 104 illustrations

Saint Louis
THE C. V. MOSBY COMPANY   1974

Printed in the United States of America

Distributed in Great Britain by Henry Kimpton, London

Library of Congress Cataloging in Publication Data

Giesel, James T   1941-
   The biology and adaptability of natural populations.

    1. Animal populations.   2. Plant populations.
3. Population genetics.   4. Ecology. I. Title.
[DNLM: 1. Adaptation, Biological.   2. Ecology.
3. Genetics, Population. QH431 G455b 1974]
QL752.G53   575.1   73-18288
ISBN 0-8016-1812-6

GW/M/M   9   8   7   6   5   4   3   2   1

PREFACE

The study of population biology is composed of two disciplines: genetics and ecology. The first of these, population genetics, deals with the way population variability is derived and maintained. It considers the effects of mutation rate, migration, population size, and chromosomal organization. These processes, interacting with natural selection, permit populations to adapt and evolve to meet the ever-changing exigencies of their environment, and they form the basis of the evolution of new species. On a theoretical level, population genetics attempts to determine how the processes just mentioned are integrated and how they may interact to enable a population to cope with a heterogeneous and changing environment. It is the purpose of theoretical population genetics to determine how these processes are integrated, how they interact with each other, and whether and how a population will be able to adapt when faced with a fluctuating environment or an environment in which conditions are changing directionally.

Population ecologists, students of the second discipline, have been concerned largely with the investigation of the variables that influence the rate of growth of populations and with fluctuation in population size in an environment in which some resource is present in quantities-limiting size. The study of systems of interacting populations, or communities, is becoming increasingly important. The dynamic interrelationships of organisms in a community is of great and pressing interest at a time when man is seriously disrupting the organization of his biotic environment.

Unfortunately, since the 1930s population genetics and population ecology have grown apart. For ease of calculation, population genetics has tended to disregard the fact that populations are usually composed of individuals of different ages with, perhaps, different rates of growth. Population ecology has historically ignored the fact that populations are composed of genetically diverse individuals. In both disciplines more realistic approaches are having significant effects on the progress and direction of population biology.

This book treats population biology in as integrated a fashion as possible and is therefore a study of adaptive or existence strategies. To some this approach may seem to limit the book's value, but careful consideration will reveal that this is really what ecology and population biology are about.

The first chapter deals with processes of population genetics. The second chapter shows the relationship between the intensity of natural selection and the population's growth rate and discusses the regulation of population density. The remainder of the book presents an analysis of the interactions of two or more species within the ecological community. At all stages of development of the community concept, the fact and implications of genetic variability within populations and between members of the community are emphasized. It is my hope that the reader of this book will gain an intuitive sense of the principles of population and community organization and evolution.

The subject matter is approached primarily through a series of examples. In all cases these examples make particular points and illustrate particular principles. It is hoped that such an approach will be interesting and maximally instructive to beginning students of population biology. Portions of this book have been used with success as supplementary reading in the introductory zoology sequence and in the undergraduate ecology course at the University of Florida. It is anticipated that the material presented and the method of its presentation will lend themselves to use in a variety of evolution and beginning population biology courses.

I thank Peter Frank and Richard C. Lewontin for guiding and inspiring my doctoral and postdoctoral years, respectively. My thanks also go to Dr. George B. Johnson, who reviewed the manuscript, and to my colleagues at the University of Florida, particularly Carmine Lanciani, for many stimulating discussions. Dr. Lanciani and Dr. William Carr allowed me to use data from their excellent but, at press time, unpublished experiments; and Dr. James Jackson, Dr. Jon Reiskind, and Miss Chris Simon provided several interesting illustrations.

But I owe my greatest debt of gratitude to my wife, Betty Jean, who criticized, edited, and typed the manuscript. Without her constant support and encouragement this would have been a far more difficult undertaking.

JAMES T. GIESEL

CONTENTS

INTRODUCTION

Since the characteristics of any population or group of organisms are functions of its environment, it is logical to begin a study of adaptational strategies with a consideration of environmental factors, such as temperature, humidity, and chemical factors. The effects of other organisms are also important, both directly in terms of competitive and predative interrelations and indirectly in terms of their effects on the physical environment. Later chapters are devoted to competition, predation, and other more complicated species interactions. This first short section concerns physical factors of the environment, their geographic pattern, and their ultimate gross effects on species and population distributions. Although ideas of microhabitat are discussed in detail later in this book, a brief introduction to this concept is given in this section. Some common precepts of animal and plant physiology are also introduced, since much of the study of population dynamics and adaptive strategies are necessarily based upon these.

CLIMATOLOGY AND GEOGRAPHY

Almost everyone is familiar with the basic way in which our planet is divided climatically: oceans and land masses are divided into tropics, more variable and temperate subtropics, temperate areas, and finally polar areas. All of these latitudinal zones have characteristic florae and faunae.

The major parameter of latitudinal zonation is temperature, which may be highly variable. It is gener-

ally low at high latitude and increasingly stable and high as the equator is neared. Differences in mean temperature do not seem to be as important floristically and faunistically as are differences in the range of temperature experienced. For example, New York City has higher extreme temperatures than do most tropic areas, and temperate North American high temperature records (134° F) greatly exceed those known for the South American tropics. These temperate regions, experiencing greater extremes of environmental factors, exhibit a far less diverse biota than do the more environmentally stable tropics.

Position of the sun

Major geographic differences in temperature are primarily caused by (1) the relation of the sun's angle to areas of the earth at different times of the year and (2) the distance of the sun from these areas at various times of the year. As shown in Fig. 1-1, the earth maintains an angle 23.5 degrees from perpendicular with the sun. As the earth circles the sun each year, the angle of solar incidence, which results from the earth's rotational angle, determines how much solar light energy will pass through different thicknesses of atmosphere at any given latitude at different times of the year. As light energy passes through the atmosphere, a proportion of it is both reflected and absorbed by water molecules, carbon dioxide, and dust. Thus increased distance of passage results in increased extinction of incident energy. For example, in June north latitude is closest to

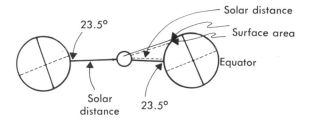

Fig. 1-1. Cause of seasonality and latitudinal temperature distributions. Incident radiation per unit area (and therefore solar heat input) are defined by $I = I_c e^{-SD n_c}$ where I is energy incident on the earth's suface per unit area, I_c is incident energy in a light band of unit area, SD is the distance the light beam travels through atmosphere, and n_c is the atmospheric extinction coefficient. The surface area, a band of unit area, impinges on increases with angle of incidence. This increase in impinged area dilutes energy input per unit area.

the sun and, therefore, receives the most insolation, but in December it is farther away and consequently cold and energy poor. Variation in temperature diminishes with approach to equatorial climes and is minimal at exactly 0° latitude, the equator.

A second function of the sun's angle that is important in determining climate is the rate of solar energy input per unit area of the earth's surface. Because a band of light of unit area must impinge on larger areas of surface as the angle of incidence decreases from perpendicular, the rate of input of incident energy per unit area must decrease. Therefore, not only is the amount of energy that reaches the ground in a polar area less than that at less extreme latitudes, but also there is less energy per unit area. Such differences result in a latitudinal distribution of climates.

Air masses

The latitudinal pattern is accentuated by the resulting movements of major air masses. To understand this, recall for a moment the physical properties of air. Air, a gas, expands when heated, lowering its density; thus warmer air rises through cooler, more dense air. The earth's air is heated by solar radiation. Because photic energy is reflected at the earth's surface and in large part is transformed into thermal energy, air near the earth's surface is generally warmer than air at higher altitudes. Equatorial air, which is warmer than air at high latitudes, has a greater tendency to rise. This upward movement of air at the equator causes more dense air from higher latitudes to move toward the equator (Fig. 1-2). This replacement phenomenon is the cause of the major winds that blow from north and south toward the equator. The warm air that rises at the equator eventually cools at high altitudes and sinks back to the earth's surface at about 30° latitude. A second wind going from 30° toward the poles is set up, again by latitudinal differences in air pressure.

The air mass that rises near the equator cools and becomes more dense as it rises, causing water vapor condensation. This phenomenon is responsible for the heavy equatorial rains. This same air, now much drier, subsequently falls at 30° latitude, warming as it falls to lower altitudes. The dry air traps moisture, resulting in major desert areas at these latitudes in both northern and southern hemispheres.

If air movements were no more complicated than this, the earth's surface would have only a latitudinal distribution of temperature, rain, and wind patterns. However, another parameter of climate, longitudinal distribution, occurs because of the earth's rotational, or Coriolis, force. Coriolis force results from the earth's axial spin and from variation in rates of movement of points located at different latitudes on the earth's surface. If we think of velocity as distance traveled per day, then it is obvious that a point at the equator has a greater velocity than points at higher latitudes (Fig. 1-3). Because the earth tends to rotate more rapidly than the atmosphere above it, the movements of air masses relative to points on the earth's surface are in an opposite direction from rotational movements of the surface points. Since the earth rotates from west to east, air masses in general move past a given point on the earth's surface from east to west.

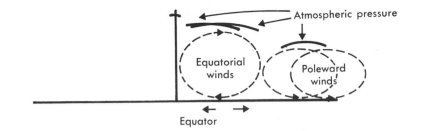

Fig. 1-2. Latitudinal patterns of prevailing winds determined by patterns of heat flux and atmospheric pressure. Pressure is a consequence of flux and vertical air motion.

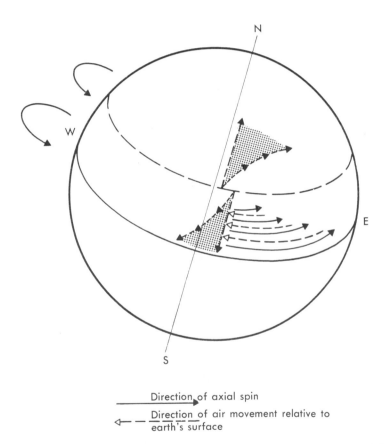

Direction of axial spin

Direction of air movement relative to earth's surface

Fig. 1-3. Major patterns of air movement are formed by interaction of latitudinal high and low pressure areas which cause major equatorward and poleward flow patterns. Patterns originate at about 30° north and south latitude. Pressure areas interact with the Coriolis force to produce winds shown by curved arrows.

Imagine the air mass moving toward the equator from 30° N latitude. It will be moving from an area of lower east-west velocity to one of higher rotational speed; as it moves nearer the equator, its east-west velocity will increase. The net result is that winds within 30° of the equator blow from northeast to southwest. Similarly, an air mass traveling north from the 30° N secondary high-pressure area is going from a region of higher rotational velocity to one of steadily decreasing velocity. Therefore, it must lose some of its east-west momentum as it moves northward. As a result, air flowing toward the north pole is deflected eastward and forms winds that blow from southwest to northeast. Such air movements, together with the location of major land and water masses, result in rainfall and climatic patterns and complete the climatic distribution.

Mountain ranges

Mountain ranges also play an important role in determining climate. For example, in Oregon (between approximately 42 and 46° N latitude) warm, moist air arises over the Pacific, sweeps eastward, and upon encountering the first of a series of coastal mountain range rises, cools, and drops its water load. Thus the western slopes of West Coast mountains are wet, and the eastern slopes lie in a rain shadow produced because falling air masses pick up rather than lose water.

This same general air mass crosses valleys lying between coastal and more inland mountains. It will lose water if the land mass can cool the air further. Generally, the interior valley of Oregon is drier than the western slopes of the mountains. Winds eventually reach the western slopes of the Cascades and Sierras, depositing more precipitation as they cool. A major rain shadow results on the eastern side of these mountains, producing the "great American desert" of Nevada and Utah. The Rocky Mountains form another major rain shadow. As a result, the Northern Great Plains are quite arid. More moisture is eventually picked up as the winds cross inland bodies of water.

It is worth noting that as the latitude nearest the sun changes with season, both the focal point and the world pattern of wind formation change. Winds in most temperate latitudes change direction with season, and this can result in quite unstable weather.

Lakes, oceans, and vegetation

Lakes and oceans temper climate. Bodies of water such as Lake Michigan act as heat traps during summer, slowly releasing their thermal load in fall. Consequently, western Michigan winters are milder than those in central Michigan. However, western Michigan springs are cooler because the lake tends to remove heat energy from air masses sweeping across it. Cooling air masses result in spring rains. Thus we have a two-dimensional pattern of climate. Plant distributions conform to this pattern, and in doing so they modify the environment, producing a third dimension, a matrix of microclimatic pockets. Such microclimatic pockets result from the effect of large-scale vegetation patterns on air movement and moisture content. For example, a desert area is hot during the day but cools rapidly after sundown. The cooling ground cools the air above it, producing a focus of high pressure, at least if the desert is surrounded by forested land, which cools more slowly. This will set up an anticyclonic pattern of winds that blow from the desert to surrounding areas. Conversely, the desert warms faster once the sun rises, and cyclonic winds result, which flow out over the desert.

Further evidence of the relationship between plant distribution and climate comes from simple observations. The western Cascade slopes are covered with Douglas fir at low elevations and with cold-tolerant pines and firs at higher altitude. Eastern slope areas are sparsely covered, first with Ponderosa pine, then with sagebrush and juniper. Similarly, the great midwestern prairies are functions of climate, but also of history. Ecologists have difficulty explaining why some areas of Illinois and Indiana were covered with grassy prairies before they were cultivated. The climate is conducive to hardwood forest species, yet grassy areas persist. Some think that unbroken prairie grass can successfully exclude tree seedlings. Others think that fire maintains grassland by continually killing trees and bushes before they can reach reproductive maturity.

BIOGEOGRAPHY

The result of a mosaic of climatic areas is a large-scale mosaic of vegetation complexes, the component organisms of which are largely determined by physical factors of their environment and by past history. Many attempts have been made to quantify climatic reasons for global patterns of vegetation structure. One of the most recent of these was devised by Holderidge. An important illustration adapted from his work is presented in Fig. 1-4. Holderidge takes into account temperature, yearly rainfall, and an aspect of plant physiology, evapotranspiration potential. The last is a function of leaf structure and, most important, temperature. Evapotranspiration increases with increased ambient temperature; thus, because of water relations, the character of vegetation changes with latitude even though the different zones have the same amount of rainfall. Southern arid areas like parts of Texas are characterized as desert, whereas northern areas with the same amount of rainfall are heavily forested. Although these forest plants have evolved the anatomical components that allow them to prosper in relatively dry habitats, their job was undoubtedly made easier by low ambient temperatures.

Plant species are often limited by water relations. Water uptake potential must at least equal evapotranspiration rates (rate of water gain = rate of loss). White ash, scarlet oak, and sugar maple are all adversely affected by decline in annual precipitation from normal levels (Fig. 1-5). Comparatively speaking, most plant species have different ranges of tolerance and different responses to climatic variation and extremes. Their distributions are thus determined partially by climatic factors. Other important factors are light intensity and soil mineral concentration. For example, hemlock seedlings fail to survive in open fields, but do well in shaded areas. Hemlock is characteristically a secondary forest succession species, which is adapted to conditions of low light. In contrast, the Douglas fir grows well in open fields and requires shade only as a very young seedling. The tree of heaven is rapidly killed by low light intensities, being adapted to life in open fields. Other species are rigidly adapted to (have tolerance limits for) life on special soil types with characteristic moisture contents, drainage characteristics, densities, and chemical composition. Similarly, animal distributions are related to climatic characteristics and to the extent to which physiological tolerance and acclimation are able to cope with environmental conditions. For example, there are latitudinally defined races and subspecies of many common animals such as cottontail rabbits and white-tailed deer.

Physiological adaptation can take many forms. For example, mechanisms of thermal acclimation differ between homeotherms (warm-blooded animals, such as mammals and birds) and poikilotherms (cold-blooded animals, such as fish, reptiles, and invertebrates). In homeotherms, adjustment to changes in ambient temperature is a whole-body process based on maintaining some constant ''core'' temperature. This adjustment may consist of growing and shedding heavy fur coats, of behavioral mechanisms such as choosing a habitat with near-optimum temperature or migrating, or of changing metabolic pathways to produce more or less body heat. Such metabolic acclimation involves hormonal, nervous, and biochemical control and modification. Often diet changes are also involved. In winter you probably eat more meats and foods high in fat content, and during hot summer months you probably eat more fruits, vegetables, and lean meats. These various foods are metabolized with different efficiencies and result in production of different amounts of metabolic heat.

Response to temperature variation is a rapid process in poikilotherms. It may involve behavioral and anatomical modifications, as in lizards (see Chapter 1),

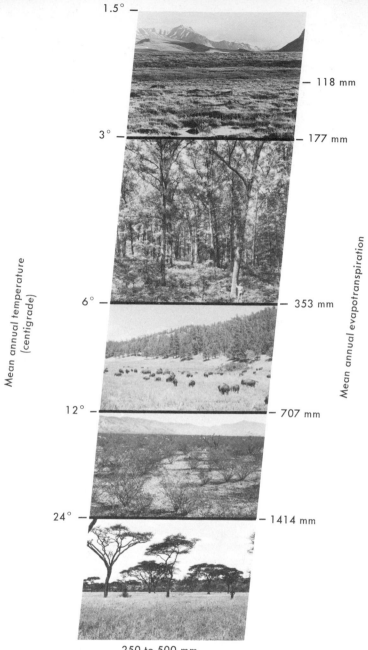

Fig. 1-4. Some major vegetation types. They are determined in general by temperature and water supply. (Photos from Kucera, C. L. 1973. The challenge of ecology. The C. V. Mosby Co., St. Louis.)

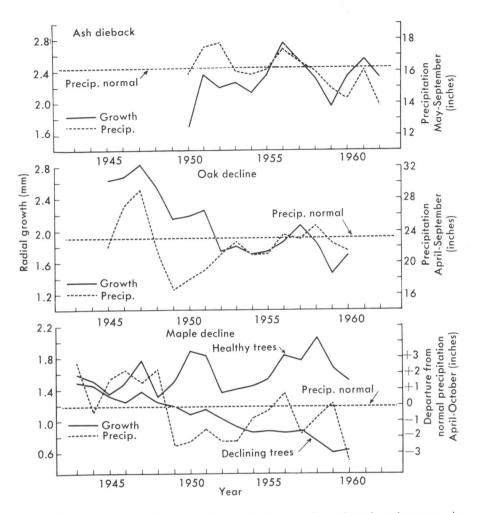

Fig. 1-5. Relation of precipitation to radial growth of white ash, scarlet oak, and sugar maple. (After Sinclair, W. A. 1964. Cornell Plantations **20**:62-67.)

or biochemical means, as in fishes and most invertebrates. Hochachka has shown that anadromous steelhead trout use one acetylcholinesterase (AChE) enzyme in their cold, oceanic environment and another when they enter warmer streams to spawn. The AChE enzyme used in the cold ocean is most efficient at low temperatures; the freshwater enzyme works best under warmer conditions. This trait, known as developmental switching, is common when changes in environmental parameters are predictable. In the temperature-sensitive enzyme systems of the steelhead, mechanisms designed to handle osmotic stress and other physiological processes are modified in preparation for the fish's periodic migrations between fresh and salt water.

In the majority of other poikilotherms, acclimation involves rapid changes of function of single enzymes and of enzyme systems that alter an array of biochemical reactions and interactions. These alterations apparently involve changes in the substrate affinities of major enzymes. Changed substrate affinities seem to compensate for changes in thermal energy of reactants, thermal energy being a direct function of the temperature of reaction systems. Although physiological mechanisms of response to a variety of factors such as temperature and pressure (see book by Hochachka and Somero) differ, their effect is roughly the same. Organisms that can acclimate well are able to exist efficiently over a wide range of environmental fluctuation and variance.

Physiological tolerance ranges, especially to temperature extremes, seem to be related to body size in both poikilotherms and homeotherms. Large animals should have more stable core, or inner body, temperatures than smaller ones, because their greater tissue mass acts as a buffer against thermal fluctuation. Partially because of this relationship between physiological tolerance and body size, distributions of large animals tend to be of the same scale as the major plant mosaic, but smaller homeotherms usually have smaller species ranges. As examples black-tailed deer range the area between the Pacific Ocean and the Cascade and Sierra summit; the mule deer, largely an arid land species, lives east of the summit. Timber wolves were once widely distributed over forested areas; snowshoe hares are limited to the northland. Nonmigratory bird species often have sharply limited distributions, as do some migratory birds during breeding season. Such limitation may be related to the presence or absence of food species, but it may also be related to the problem of shelter. Lizards and other cold-blooded animals, as well as insects, are found only in the shelter of vegetation or under logs and other debris, which moderate ambient conditions; these animals may be prey for some habitat-limited birds.

Green plants and other substrate-covering organisms (see Chapter 5) and materials alter the physical environment in their immediate vicinity and produce microhabitat patches, each with its own characteristic climate. Within such microclimates physical factors such as temperature, humidity, and wind velocity are moderated and generally are far less variable than they are in the environment at large. Table 1-1 shows examples of the effects of vegetation on temperature and humidity. Ranges of variation for these climatic factors are compared for a plowed field, grassland, and forest at a variety of heights above ground and at different depths within the soil. Temperature

TABLE 1-1. Daily temperature variation (in degrees centigrade) during August in Austria*

	ALTITUDE (METERS)									
COVER TYPE	10	5	3	2.5	1.5	0.2	−0.01	−0.05	−0.1	−0.2
Forest	16.4°	19.4°	19°	18.4°	16.5°	14°	—	—	—	—
Moorland	—	—	—	—	—	—	17.3°	14.1°	8.5°	2.6°
Cut meadow	—	—	—	—	—	—	15.7°	6.8°	4.3°	1.7°
Moor with turf removed	—	—	—	—	—	—	13.1°	8.6°	4.0°	1.4°
Moor (and meadow)	—	—	—	—	—	—	7.3°	5.5°	2.9°	.8°

*Data for grasslands is for temperature variation (°C) below ground; that for the fir plantation gives above-ground levels. Temperature is less variable just above ground level in the fir plantation than just below ground in the grassy and shrubby areas. (Data from Geiger, 1965. The climate near the ground. Harvard University Press, Cambridge, Mass.)

variations in the plowed field are less with depth because of the blanketing effect of the soil, and different types of soil have different moderating capacities (e. g. sandy soil is less effective than clay). Similarly, the vegetation of grasslands and forests acts to curtail variations in temperature above the soil's surface and to raise the zone of minimum variation above ground level. This modification of the environment is biologically significant because life below ground confers unique problems in population dynamics and mating. Generally these problems are less severe for supraterrestrial species. Thus population densities and numbers of species of animals such as litter insects are higher in forests and other vegetated areas than in less physically stable areas. (This argument presupposes that food supplies are everywhere the same, which is obviously not true, and that the level of physical heterogeneity of the environment is the only critical factor determining population density and species diversity. As you will see in Chapter 5, this set of suppositions is far too simple.)

Not only does vegetation cover reduce temporal variation of humidity within a site, but it also raises average humidity (Fig. 1-6). This occurs because plants release water to the environment as a product of transpiration, a process resulting from their mode of nutrient transport. In addition, rate of heat loss of wet air is lower than that of dry air; therefore, additional temperature stability is conferred by elevated microenvironmental humidity. Stable humid microclimates are helpful to seedling germination, further plant growth, and nutrient turnover.

The effect of elevated humidity on animals, reduced water loss, is obvious, particularly in otherwise dry areas. Also, high microenvironmental humidity affects feeding relations within an insect community. Many herbivorous forest insects consume the bacteria and fungi of decay rather than living plant material (see Chapter 5). Humid climates near the ground are conducive to growth of these decomposer fungi and bacteria which attack plants after death.

Microclimates formed by vegetation are important units in the habitat mosaic, especially for smaller animals, and often for larger ones that are dependent on them. Microhabitats are of extreme importance to the biology of populations. Examples of microclimate formation are, of course, almost innumerable. The area under or on top of every leaf of every different kind of plant is a slightly different habitat. All such areas have temperatures and humidities that differ from the surrounding environment. Such small spaces serve as homes for a variety of small animals.

Along the oceans' edges, the macroscopic algae, mussels, and barnacles that cover rock faces form areas protected from heavy wave wash, fresh water, and desiccation that occurs at low tide. These microhabitats serve as homes for a plethora of crabs, snails, sponges, nemertean and polychaete worms, and other small organisms. The mussel beds not only form sheltered places for small organisms, but also provide nutrients. Mussels capture food from the open water,

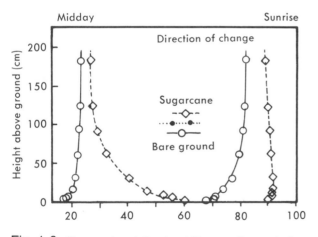

Fig. 1-6. Changes in relative humidity near Poona, India during the course of a day. The humidity near the ground and that under cover of sugarcane are both higher and less variable than that under more exposed conditions. (Data after Ia Ramadas, R. J. Kalamkar, and K. M. Gadre. 1934, 1935. Agricultural studies in microclimatology. Ind. J. Agri. Sci. **4**:451-467; **5**:1-11.)

concentrate it, and release it to the intertidal community as pseudofeces. This material is used directly by the invertebrates of mussel beds.

Although the subject of environment is not treated further in this text as a separate concept, many examples of its relationship to adaptive strategies will be given throughout the following chapters.

BIBLIOGRAPHY

Geiger, R. 1965. The climate near the ground. Harvard University Press, Cambridge, Mass.
Hochachka, P. W., and G. N. Somero. 1973. Strategies of biochemical adaptation. W. B. Saunders Co., Philadelphia.
Holdridge, L. R. 1967. Life zone ecology. Tropical Science Center, San José, Costa Rica.
Sinclair, W. A. 1964. Comparisons of recent declines of white ash, oaks, and sugar maple in Northern woodlands. Cornell Plantations **20:**62-67.

2 EVOLUTION AND POPULATION GENETICS

THE POPULATION CONCEPT

To the population biologist a population is a group of organisms, always of the same species, that interbreed and form viable offspring. This interbreeding is one of the most important attributes of natural populations. Populations can be of almost any size, ranging from the very small panmictic (randomly intermating) unit, which is isolated in some way from other groups of the same species, to very large assemblages existing over wide geographic areas and interbreeding through a series of generations. No more concrete definition of population is possible. In practice, defining the bounds of natural populations and, therefore, the size of the panmictic unit, is largely arbitrary. However, it is practical to discuss factors that have been shown to differentiate groups (populations) within a species, although even these bounds are subject to forces arising from within the group itself.

Species can be divided into subgroups according to various barriers to interbreeding. At the genetic level this might include barriers to interbreeding that stem from incompatibilities of chromosome arrangements. However, this class of isolating mechanism will not be treated further, because it is not of general importance to the biology of natural populations.

A major barrier to interbreeding is geographic isolation. Geographic and geological features of the environment often define limits of populations of a species. For example, birds within a species of the genus *Geospiza,* which occur in the Galapagos Islands, ex-hibit slight differences in morphologic characteristics and feeding behavior from island to island. These differences no doubt occur because the islands are separated by large bodies of water, which prevent intermingling of birds from different island populations, thus allowing differential selection to act on each of the isolated groups. The genetic continuity of the individual island populations and the differences among these populations are thereby maintained by geographic isolation. Mainland rivers and mountain ranges can also serve as isolating barriers.

The strength and scale of geographic isolation may vary greatly. Cultivated fields often serve as barriers to the interbreeding of insects that live in hedgerows or woodland on either side of a field. In fact, at least for short periods of time, populations of soil arthropods specific to the leaf litter of a particular species of tree may be isolated from each other simply because that species of tree is scattered throughout a forest.

If separation has not been of sufficient duration for genetic incompatibilities to have evolved among populations, members of geographically isolated populations will interbreed when the barriers are removed. For example, there are distinguishable eastern and western populations of flickers whose differentiation is partially the result of their separation by the Rocky Mountains and the once nearly treeless Great Plains. However, trees planted at farmsteads on the plains have extended eastern and western habitat across the plains, allowing hybridization of some closely related species. Thus the amount of migration and interbreeding among populations of a species varies according to the geographic and temporal situation.

Migration and interbreeding also vary according to the temporal situation. In some species interpopulation migration of high intensity occurs seasonally, and in most species migration increases as crowding increases within populations. Populations are thus isolated one from another to varying degrees, and isolation is often transient.

Isolation by distance

Theoretically, isolation of subpopulations does not require habitat discontinuity. Distance alone can cause genetic subdivision of a species. Sewall Wright recognized this possibility when he derived his island model and later the stepping stone model of population structure. It can be shown that even though interpopulation mating can occur among subgroups located at distances from each other, the probability of successful gene flow decreases with the distance that a propagule (fertile female or male) must travel to reach a new site. The process can be visualized as a continuum over distance or as a series of population centers, neighborhoods, or stepping stones. A migrant starting at one end will most likely move to the next stepping stone and is most unlikely to travel over the whole continuum. Therefore, those populations at the opposite ends of a spatial continuum are expected to be most distinct. Adjacent populations should be most similar to each other.

The size of a neighborhood depends on several factors: individual vagility, mating structure, and the species' physiological adaptability and reproductive ability in a variety of different environments. These parameters are species specific and depend on evolved strategies. Neighborhood size is also affected by the environmental variability that dispersing individuals encounter during interpopulation migration. Species from environments in which temporal heterogeneity exceeds spatial variability should have large neighborhoods, because each local population experiences temperature extremes greater than those encountered geographically. One example of such an environment is the high intertidal zone where various marine invertebrates live on exposed rock surfaces. The importance of this parameter may also be seen along the Florida coastline. Temperature variation at Miami is greater during the month of March than the difference in average during that month from Jacksonville to Miami, a distance of about 300 miles. Members of a barnacle species occurring along this section of coast seem to exhibit intense interpopulation migration. In contrast, woodland insect species might have much smaller neighborhoods because spatial heterogeneity may be more restrictive. They could show stronger isolation by distance.

Species population size is also important. Numerically large species should be more continuously distributed than small ones.

In the absence of geographic barriers, or other isolating effects, animals, including man, tend to exclude outsiders from their populations. For example, many of the Indian tribes of the southwestern United States and of South America have kept themselves at least partially separated for long periods of time. Perhaps this artificial isolation has some redeeming social or political value, but such inbreeding can have deleterious genetic consequences. In order to counter the effects of interbreeding, extremely detailed and stringent mating rules have evolved within these cultures, but such social rules are often transient. Captives taken during intertribal warfare were often assimilated into the capturing population, and this assimilation constitutes sporadic breakdown of social isolation. Thus it is not easy to precisely define a population and its bounds. Some migration almost always occurs between populations, and this migration is the final determiner of the size of the interbreeding population.

ALLELE FREQUENCY

Evolution, the adaptation of populations to new or changing environments, consists of the numerical ascendance of a superior genetic "type" within the population. It is convenient to discuss evolution in terms of changes in genotype and allele frequency. Before discussing allele frequency, however, it should be emphasized that no two organisms of a population are ever genetically (or even phenotypically) identical, except on casual observation. Many members of a population might possess the same genotype or alleles at a particular genetic locus, but every organism possesses a vast array of genes, by conservative estimate more than 10,000. Recently, it has been shown that in the average population from 20 to 84% of these loci may be polymorphic (have more than one genetic variant, or allele). If as many as half of the loci in a population have only two alleles segregating (many loci have more than two alleles) and if the two alleles are equally common, then, according to the laws of probability, the population can consist of 3^{5000} different genotypes. However, factors concerning the genetic architecture (discussed later) of individuals greatly reduce the amount of actual genetic variability. Even so, these calculations suggest that it is almost infinitely unlikely that any two individuals in a population, no matter how large the population, will be of exactly the same genotype. (Of course, organisms that are clonal, or produce vegetatively, are not included in this statement.)

Because of this complexity, it is impossible to treat explicitly the genetics of a population. There are simply too many loci that interact in too many possible ways for us to make any sense of a population's genetics at a level that considers all genes. To add further complication, the unit of selection is not really the gene, as existing theory assumes; rather it is the phenotype, the production of which is mediated by the expression of the entire genotype of an individual.

However, by treating single loci independently we can derive a rather simplified theory of population variation and natural selection. This theory leads to many important logical conclusions about evolutionary processes. In order to understand this basic theory we must first define the concept of allele frequency, often called "gene frequency."

The best way to illustrate allele frequency is by specific example. Suppose, as in Table 2-1, that in a total population of 500 Greenland Eskimos, 125 are of the MM blood group, 250 carry both the M and N alleles, and 125 are homozygous for the N allele. The frequency, or proportion, of MM individuals in the population is simply the numerical representation of this genotype divided by the total population size (125/500, or 25%). The frequencies of the MN heterozygotes and of the NN homozygotes are derived similarly ($MN = 250/500$; $NN = 125/500$, or 25%).

In order to describe population genetic processes that may be occurring in the population (such as natural selection) we need both an estimate of allele frequencies in the population and an idea of the genotype frequencies that we might expect of that population, making several basic assumptions. We can derive estimates of the deviation of observed frequencies from those expected. Initial estimates of genotype frequencies are of limited use by themselves, since we are really interested in measuring changes in gene frequency that occur within and between generations of the population. Different patterns of change may suggest particular population genetic causes.

Estimation of expected genotype frequency in the population and of the genotype frequencies expected of the next generation (a necessity for population genetic analysis) can be accomplished in two steps. First, the frequencies of the alleles in question are estimated (see Table 2-1): Because the MM homozygote can produce only gametes carrying the M allele, its contribution of M gametes to the next generation can be estimated by its frequency. The heterozygotes, MN, can produce two kinds of gametes, M and N; these will usually be produced in equal frequency, or with equal probability. Thus only half of the gametes produced by the heterozygotes will carry the M allele. The contribution of M gametes by the heterozygotes to the next generation must therefore be equal to half of the frequency of the heterozygotes in the parental population. Therefore, the expectation is that p, the frequency of M, = (frequency of MM) + (½ frequency of MN). Since any set of probabilities, if inclusive, must total 1, q, the frequency of N, = $1 - p$.

TABLE 2-1. Calculation of allele and genotype frequencies

GENOTYPE	MM	MN	NN	TOTAL
Number	125	250	125	500
Frequency	125/500 =	250/500 =	125/500 =	1.00
generation 1	0.25	0.50	0.25	
Probability production of M	1.00	0.50	0	—
Production of *M*	0.25	0.25	0	0.50
(frequency $\times$ probability) *(p)*				
Frequency				
generation 2	$p \times p$	$(pq) + (qp)$	$q \times q^*$	
	p^2	$2\,pq$	q^2	
	$(0.5)^2$	$2(0.25)$	$(0.5)^2$	
	0.25	0.50	0.25	1.00

*Probability N $= 1 - p$
$= q$
$= 1 - 0.50$
$= 0.50$

In order to estimate the genotype frequencies expected of the next generation, we imagine that the parental population has produced a pool of gametes, or alleles. One portion of these, *p,* are of one allelic form (in this case *M*) and another portion, *q,* are of the alternative allele *(N).* The gametes in this pool are assumed to unite at random to form the young of the next generation.

We can derive expected proportions of genotypes formed by this population by applying a basic theorem of probability which states: *The probability of co-occurrence of any two independent events is given by the product of their separate probabilities of occurrence.* Allele frequencies are the probabilities of occurrence of the alleles in the population. Thus the probability, or expected frequency, of *MM* progeny in the Eskimo population is $p \times p,$ or $p^2.$ Similarly, the expected frequency of *NN* progeny in the population is $q \times q,$ or $q^2.$ The probability of *MN* offspring is $(p \times q) + (q \times p),$ or 2 *pg,* since there are two ways in which $M \times N$ can occur; *M* (or *N*) can come from either a male or a female parent. This is the Hardy-Weinberg relationship: Given that a fraction, *p,* of alleles are *M* and a fraction, *q,* of the alleles are *N* in a large popula-

tion that is randomly breeding, then a fraction, $p^2,$ of the progeny will have genotype *MM;* a fraction, 2 *pq,* of the progeny will have genotype *MN;* and a fraction, $q^2,$ of the progeny will have genotype *NN.* Calculations are shown in Table 2-1.

Table 2-1 shows the mechanics of calculating allele frequency and illustrates the application of such calculations. The genotype frequencies in the progeny generation are the same as those in the parental generation. Hardy and Weinberg first made this observation independently in 1903. From their work has developed the Hardy-Weinberg principle, which states: *In the absence of disrupting factors (mutation, migration, sampling effect, selection, and linkage) the allele and genotype frequencies at any locus in a panmictic population will be repeated faithfully from generation to generation; should the frequencies be perturbed for any reason, they will come to the expected equilibrium values after one generation of random mating.*

A little thought suggests that we can turn the principle around and attempt to use it as an analytic tool. We know that allele and genotype frequencies are supposed to repeat and that predicted frequencies should agree with those observed. Therefore, if we observe

TABLE 2-2. Frequencies of shell color among limpets (trait is assumed to be controlled by a single locus with no dominance)

FREQUENCY	WHITE	TAN	BROWN
Observed frequency	0.14	0.76	0.10
p (white) $= 0.14 + 0.5\,(.76)$			
$\quad\quad\quad\quad = 0.14 + 0.38$			
$\quad\quad\quad\quad = 0.52$			
$q = 1 - p = 0.48$			
Expected frequencies (f_1)	p^2	$2pq$	q^2
p' white $= 0.2704 + 0.2496$	0.2704	0.4992	0.2304
$\quad\quad\quad = 0.52;$			
q (brown) $= 0.48$			
Expected frequencies (f_2)	0.2704	0.4992	0.2304

frequencies differing from those expected, we should, depending on the pattern of differences, be able to assign their most probable cause (assuming that we have some knowledge of how various processes such as selection and migration affect genotype frequencies in a population). We should at least be able to tell if the genotype frequencies of a population are changing in some systematic way without going to the trouble of estimating them for each generation.

However, assigning cause is not easy. For example, in the data shown in Table 2-2, first progeny generation genotype frequencies differ from those of the parental generation. There is an excess of heterozygotes over the proportion expected. The application of some factors causing excess death (differential selection) of both classes of homozygote is the most likely explanation for the observed perturbation of gene frequencies. However, in this example there is insufficient information to analyze the processes that are occurring. Another explanation might be differential migration from, or into, the population. Also, although the viabilities of the homozygotes might be lower than those of the heterozygotes, we have no information about the relative reproductive efficiencies of the three genotypes. Thus deviation of genotype frequencies from those expected, when calculated from the data of a single generation, can only suggest the most fruitful direction of future research.

When the calculation of expected frequencies is repeated for another generation (Table 2-2), genotype frequencies cease to change (they reach equilibrium values) after one generation of random mating. This illustrates the Hardy-Weinberg Law.*

The basic probability-generating machinery suggested in the Hardy-Weinberg Law and its assumptions have been the basis of investigations in population genetics for over 50 years. The idea of allelic and genotypic probability is a powerful basic tool in itself. The processes of mutation, migration, and selection, and the effects of small population size, which can modify the distribution of genotype frequencies, are the nuclei of most of the concepts that are of interest to population biologists.

Discussion of these processes is the basis for the remainder of this book, since all aspects pertaining to

---

*The sex-linked gene is an exception to the rule. In this case two thirds of the genes at the locus are carried by females, while one third are on the male's X chromosome. If the gene frequencies among males are not the same as those among females, the population cannot strictly be said to be in equilibrium. As a sex-linked gene approaches equilibrium frequency, its distribution between the two sexes oscillates. The oscillations are of decreasing amplitude, and eventually a stable state is attained. The difference in frequency between the two sexes is halved at each generation, and a large number of generations may pass before true equilibrium is reached.

regulation and interaction of species populations can be related to them. Studies of the genetics and the ecology of populations are logically inseparable.

POPULATION GENETICS—GENETIC VARIABILITY

The keys to the persistence of any natural population are its genetic variability and its ability to withstand environmental change. The place of genetic variability in the repertoire of responses of a population to environmental variation and stress is shown schematically in Fig. 2-1. The figure consists of an arrow, narrow at one end and broadening toward the other end. The breadth of the arrow represents the degree of environmental stress to which an organism is exposed; although not every individual of a population responds in the same way, the responses of organisms exposed to stress can be categorized according to the intensity of the stress.

The most common response of an organism to a low level of stress is behavioral. For example, an animal's behavioral response to higher than normal temperature usually involves seeking shade or retreating to cool

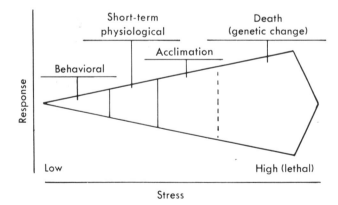

Fig. 2-1. Single organism responses to environmental stress. Larger areas within segments of the arrow represent higher stress. At the death point is the basis for evolution, since death will affect some genotypes (and phenotypes) more than others.

burrows in an effort to escape the heat. Many species of desert lizards do a remarkable job of behavioral regulation of body temperature. Since these animals are poikilotherms, they are largely incapable of regulating body temperature by physiological means; they must maintain their body temperatures at or near an optimal level by behavioral or anatomical means. A lizard, peeking from its burrow in the cold early morning will expose only a small proportion of its body to the sun's heat until it has absorbed enough heat via its exposed body surface to raise the temperature of its entire body to a level permitting normal activity. A cold lizard moves very slowly. If it attempted to begin normal foraging activities before warming up, it would be easy prey for predators. The lizard feeds in direct sunlight until its body temperature begins to exceed optimal levels. Once optimal body temperatures have been exceeded, the lizard begins to seek shade and returns eventually to its cool burrow.

If the level of stress goes beyond the point where behavioral responses are adequate, homeothermic animals will respond physiologically. For example, a dog will pant at an increased rate to dissipate excess heat. Some homeotherms greatly reduce their metabolic activity and enter a state of deep sleep or torpor in burrows, thereby avoiding the source of stress. Lizards and other poikilotherms have little such recourse to physiological means for meeting heat stress.

Beyond this second level are intensities of stress that require deepseated physiological responses. Animals will acclimate to a highly stressful evironment of long duration by modifying diet or by utilizing food reserves that can be metabolized to minimize metabolic heat production. During warm weather fur-bearing animals lose their heavy coats in favor of much thinner body coverings; they regain their heavier coats at the approach of cold weather.

Predictable stress of long duration is necessary for acclimation, since the physiological processes involved are often extremely complicated and require time to develop. (See Hochachka and Somero for discussions of physiological-genetic means of acclimation.) Short-term, high-intensity stress is fatal to many organisms. This type of stress is represented in Fig. 2-1 by the part of the arrow of greatest width.

Various environmental parameters have been determined for a wide variety of animals in terms of the LD_{50}, which is the level of stress at which one half of a laboratory population dies. Most physiologists have found more fruitful approaches, but this work contains a valuable lesson for the population biologist. Animals vary in the amount of stress they are capable of sustaining. Only part of a population succumbs to any particular level of stress. This variation is shown in Fig. 2-2. Any of the bell-shaped distributions shown represent the distribution of phenotypic expression of a genotype at one genetic locus.

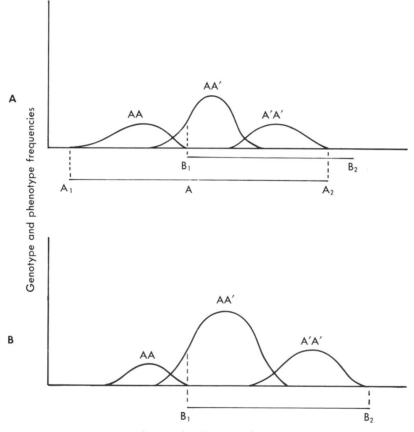

Fig. 2-2. Genotype frequency change in response to selection in a changing environment for one major gene. Each genotype produces a distribution of phenotypes because of action of modifier genes, physiological response, and environment. A preselection set of genotypes and phenotypes is shown in **A.** A possible result of the change is shown in **B.**

There are often several different phenotypic expressions of a single genotype because no two individuals are alike at all of their genetic loci. In addition, the expression of a particular locus is influenced either by that gene's intrinsic environment or by the effect of products of the constellation of other genes present within the individual. For example, the kinetic behavior of an enzyme (the product of one or a few interacting genes) is affected by, among other factors, the pH and organic content of the cell, which are affected by the products of the rest of the cell's biochemical activities. The level and proportion of these products are in turn controlled by the total genotype of the cell. *The expressions of most of the genetic traits of an organism are functions of the expressions of the genes that comprise the rest of the genome of that organism. These determine the gene's intrinisic environment.*

A population composed of only one genotype at a particular locus will be capable of only limited genetic or evolutionary change in response to a severe change in the environment. (Since the major locus is invariant, the population can evolve only within the limits of the constellation of modifier genes.) In contrast, a population composed of individuals of several genotypes (as shown in Fig. 2-2, *A*), each with its own range of phenotypic expression, is in a position to change genetically and thereby to persist numerically in response to a change in the environment. For example, imagine that Fig. 2-2 represents a population of small arthropods that live in a leaf litter of the forest floor. Most members of the population can survive only in the vicinity of environmental temperature A, while some can live well at temperatures on either side of this optimum. Let us assume that each animal can survive and reproduce only at temperatures very close to its optimum; i.e., the range of physiological acclimation of the animals is small. Let us further assume that during a relatively short period of time the mean ambient temperature of the leaf litter is represented by point A, and that this is the temperature half way through the depth of the leaf litter. Since the animals comprising the

population are expected to occur throughout the leaf litter, the range of temperatures available to the population is that between temperatures A_1 and A_2. (Temperature and other enviromental parameters, such as humidity and food particle size, vary with depth of the litter; during a summer day higher temperatures prevail near the surface of the litter, while deeper in the substrate the temperature is lower.) Under the temperature regimes encompassed by A_1 and A_2, all members of the population can survive and reproduce; the range of environments includes the optima for all animals of the population.

Now imagine that environmental temperatures suddenly increase. In Fig. 2-2, *A,* the new environment is represented by the range $B_1 - B_2$. The part of the population having temperature requirements outside the new environmental range can no longer survive and reproduce. Thus these animals will contribute no gametes to the next generation. Fig. 2-2, *B,* shows the expected result of this selective death and failure of gamete contribution. Nearly 100% of the *AA* individuals and 20% of the *AA'* individuals (of the population in Fig. 2-2, *A*) do not contribute to the next generation, (shown in Fig. 2-2, *B*). As a result, the proportional distribution of genotypes in the progeny generation differs from that of the parents. The population is adapting to its new environment. Note that population size need not be any smaller than it was before the environmental modification: it is the representation of genotypes that changes. During successive generations, lesser proportions of the population are of the unfit genotypes, resulting in less selective death. It should be obvious that *a monomorphic AA population probably would have become extinct because of the changed conditions. The polymorphic population is adapting to meet the change.*

During this process new variability is released, since the genomes of a population change from generation to generation because of recombination of loci, and different genomes are phenotypically different. Eventually, as a result of selection, of recombination, and perhaps of mutation, the population will fill most or all of the environmental space represented by the new set of temperatures. Remember that the population persisted long enough to evolve *only* because it was genetically variable initially. Variability is one of the important keys to the evolution and persistence of natural populations, all of which exist in environments that are constantly changing.

Mutation

There are various processes that ensure the genetic variability of a population. The first of these is mutation. All genetic variability at the single gene level arises ultimately from point mutation; that is, from the substitution, deletion, or addition of a single base in the code carried in the cell's DNA for an enzyme or structural protein. When the DNA of a gamete mutates, the mutant form may become a part of the next generation and thus of the genetic constitution of the population.

Dominance. Most mutant forms are phenotypically recessive to the allelic form from which they arose. It is thought that a mutant is initially recessive because the extent of phenotypic display of the enzyme product of an allele is a function of that enzyme's associations in the cell's enzymic structure. A functionally new form of enzyme would not be as well integrated into the cell's function as was its predecessor and, therefore, it will tend to be less efficient and to produce less than its predecessor. Thus the mutant allele appears phenotypically recessive.

Evolution of dominance. Although the genomic complex argument may define part of the reason for the evolution of dominance of alleles, R. A. Fisher developed a much more rigorous theory. According to this theory, if homozygous genotypes at a locus have different fitness values, selection should favor the evolution of heterozygotes that are phenotypically similar to the more fit homozygote. Fisher's theory assumes the presence of modifier genes that tend to enhance the functional effect of the favorable allele or detract from the functional effect of the deleterious allele in heterozygotes. Several workers have criticized this theory, suggesting that once a major gene function is determined, the action of modifiers may not be selectively significant. However, pleiotropic genes (which have more than one function) are almost always dominant for their beneficial effects and recessive for their deleterious effects. This fact seems to offer impelling evidence in favor of Fisher's theory. Most mutations are phenotypic recessives, and for reasons discussed under selection (see p. 32), most such mutants are preserved in the gene pool of the population. The concepts of gene interaction and coadapted gene complexes have been implied in this discussion. They will be treated in some depth later in this chapter.

Selective significance of mutations. Because point mutations result in changes in the structure of enzymes, a mutant enzyme will usually be functionally different from its predecessor. For example, any change in charge distribution at the active catalytic site of an enzyme or at its substrate binding site may change the relationship of the enzyme to its substrate. This, in turn, results in changed reaction rates. If an amino acid near the substrate binding site of an enzyme were changed so that it carried a net negative charge rather than a net positive charge, the substrate affinity and resulting kinetics of the mutant enzyme could differ from those of the original form. The new enzyme might bind substrate more or less readily; speed of substrate release might also be different. Some point mutations result in changes in enzyme function of a sufficient extent to be physiologically and selectively significant. Most, however, may not involve significant changes in function and physiology and may thus be selectively nearly neutral.

Most population biologists believe that the majority of mutations are selectively significant. However, a group of investigators, including Motoo Kimura, have advanced the hypothesis that a significant proportion of mutations produce alleles that are selectively neutral. Much of this argument is based on the idea that random substitutions of certain sets of amino acids in an enzyme should produce no functional differences in the enzymes. This hypothesis has received some support in theoretical and laboratory studies and in statistical studies of populations, which indicate that variations in frequency of electrophoresable proteins can be explained by a model involving interaction between random changes due to the effect of small population size (see p. 25) and known rates of gene mutation.

Despite such support for the neutral allele hypothesis, a growing body of evidence from studies of enzyme polymorphism in natural populations rather strongly refutes it. Many population genetic studies have produced evidence of the overriding importance of balanced selection in maintenance of polymorphism.

Of the functionally significant point mutations, most are likely to be deleterious. The genome, or total genetic constitution, of an organism living in a relatively long-term, stable environment is likely to be highly coadapted. This suggests that natural selection in a stable environment will lead to establishment of populations whose members possess constellations of genetic forms in which the enzymic products are highly integrated functionally. Selection will be for establishment of enzyme systems that have compatible temperature, pH, ionic, and substrate-product concentration kinetics. It is very likely that a disturbance of function resulting from the presence of a mutant allele will be detrimental to the system if the environment remains constant. Therefore, the majority of mutations would be expected to be deleterious, and this is what is observed.

Work done in the early 1950s by Morton and co-workers illustrates the incidence of lethal mutation in humans. Table 2-3 shows that mortality increases with degree of consanguinity of the marriage or with the extent to which the parents are lineally related. Recessive lethal mutants, such as the more than 300 known to exist in humans, would give these results. On the average each person carries six to eight deleterious mutants.

TABLE 2-3. Incidence of infant and child mortality as related to degree of interrelatedness of parents*

RELATIONSHIP	STUDY A		STUDY B		STUDY C
	STILLBIRTHS AND NEONATAL DEATHS (PROPORTION)	INFANTILE AND JUV. DEATHS (PROPORTION)	STILLBIRTHS AND NEONATAL DEATHS (PROPORTION)	INFANTILE AND JUV. DEATHS (PROPORTION)	CHILD MORTALITY (PROPORTION)
First cousins	0.111	0.156	0.064	0.121	0.229
Second cousins	0.074	0.113	0.046	0.074	—
Double first cousins	—	—	—	—	0.537
Third cousins	—	—	—	—	0.136
Uncle-niece	—	—	—	—	0.434
Not related	0.044	0.089	0.032	0.056	0.160

*Data in A and B from Morton, N. E., J. Crow, and H. J. Muller. 1956. An estimate of the mutational damage in man from data on consanguineous marriages. PNAS **45**:855-863. Data in Study C from Bemiss, quoted in Morton et al.

As will be shown later, it is almost impossible to remove such defects and diseases from the population; but techniques exist that aid us in avoiding their social and economic consequences. In cases where the familial history of prospective parents indicates that their future offspring might have some genetic defect, the couple can be advised to avoid having children, but since in most cases there is the possibility that the offspring will be normal, parents often assume the risk of having an abnormal child. When this occurs, small samples of amniotic fluid may be taken and analyzed by protein electrophoresis, which allows us to detect the presence of some abnormal enzymes long before the baby is to be born. Other enzyme assays can be done on biopsy specimens and the mother's blood. Thus the necessity of treatment can be anticipated and can begin at birth; if no treatment is known, therapeutic abortion can be suggested.

Such prenatal screening for deleterious mutants is likely to become even more important in the near future. There are probably deficiency diseases that have not yet been genetically characterized, and the appearance of new mutants is becoming more likely because of increasing levels of mutagenic ionizing radiation and of potentially mutagenic environmental pollutants of chemical nature.

This discussion of the effect of mutation is based on rather scanty information. We have little idea how well integrated the genome really is. Richard Levins has suggested that rather high mutation rates should be beneficial to populations in fluctuating environments, since such populations are continually faced with new conditions and might be able to use new mutations immediately at their inception or to use existing ones at some unpredictable time. According to Levins' theory, the mutation rate should increase as unpredictable selectively significant, environmental variation increases. (Levins' definition of selectively significant variation is explained on p. 35.)

The effect of point mutation on gene frequency. The Hardy-Weinberg principle explicitly suggests that point mutation will cause a deviation of gene frequen-

cies from those expected. Let us investigate this contention, assuming (1) that mutation is a random event, which may occur as a base change anywhere along the DNA strand coding for a particular enzyme, and (2) that such mutations result in the conversion of genes from one form to another. In the following algebraic analysis, it is assumed that both the "wildtype" and "mutant" forms of the gene are present at some initial frequency in the natural population. These frequencies are represented by $p(a)$ and $q(a')$, where $q(a')$ is the frequency of the mutant form of the gene and $p(a)$ is the frequency of the wild type allele. The rate of mutation from a to a' (the proportion of a genes that mutate to the a' form per generation) is designated as u. The results of mutation, in terms of change in gene frequency per generation can be written as:

$$q_1(a') = q_0(a') + p(a)u$$

q_0 is the initial frequency of the mutant allele, and q_1 is its frequency after one generation of mutation and random mating. This equation states that the frequency of the mutant allele will be augmented in each generation by $p(a)u$, the product of the mutation rate and the proportion of the wild type allele in the population available for mutation.

Just as we expect gene mutation from wild type alleles to mutant forms, we might also expect reverse mutation—a change in gene function from the mutant form back to the original wild type. Mathematically, this becomes: $q_1(a') = q_0(a') + p(a)u - q_0(a')v$, where v is the rate of back mutation. The frequency of the mutant form is augmented by an amount $p(a)u$ at the same time it is decreased by an amount $q_0(a')v$. Algebraic manipulation of the relationship allows us to predict that there will be an equilibrium allele frequency, $\hat{p}$, reached by the population. The equation $\hat{p}(a) = \dfrac{v}{v + u}$ states that the population should eventually reach an allele frequency that is stable and that depends only on the relative rates of forward and back mutation at the locus of concern. This sort of treatment suggests that mutation should have profound effect on allele frequencies; if all other factors are unimportant, mutation should, in fact, *determine* allele frequencies.

However, there are a number of problems with this analysis. First, mutation rates are commonly very small. The rate of mutation, of course, depends to some extent on the size of the gene; large genes are more liable to base substitution, loss, or addition than are shorter blocks of DNA. However, in general, mutation rates are of the order of 10^{-5} to 10^{-6}; that is, one mutation is expected to occur on the average for every 100,000 or 1,000,000 gene replications. The probability of back mutation is lower by a factor of 10 or so, because the possible changes in coding necessary to produce a *particular* reversion of enzyme function are more constrained than are those giving the original change. Considering all of the other factors that influence allele frequencies in natural populations, the influence of mutation is actually small.

Although point mutation has almost no effect on allele frequencies in natural populations, it does contribute to a population's genetic variability. As has been stated, some mutations are probably neutral at their inception and will become incorporated (at low frequency) into the gene pool of a large population. In small populations some of these mutants will be lost because of random drift (see p. 27). Other mutants will be deleterious under the environmental conditions of their inception, but they are also likely to be phenotypically recessive and will be maintained as heterozygotes in the population at low frequency. These mutants confer initial variability to the natural population. This variability is available for use should environmental conditions change.

An occasional mutant is advantageous at its inception either because it is superior in the present environment or because it arose simultaneously with some major change in the environment. The immunity of mosquitoes to DDT is the result of a mutant enzyme that denatures the pesticide. Whether fortuitous mutations are of value to a population depends on the relationship between the intensity of selection and the species' birth rate. Species with high birth rates can often survive drastic changes in their environment, but those with lower birth rates are not able to withstand high intensities of selection, and they become extinct. These interactions will be discussed later in the chapter.

Chromosomal mutations. Heritable changes in chromosome number or structure can be considered mutations. Those having greatest implication for students of population genetics are chromosome breakage, polyploidy, chromosome segment inversion, and translocation.

Chromosome breakage, when the segments can be recovered, is important because of genetic linkage. If two genes are located on the same chromosome near each other, their alleles will tend to segregate with each other. Loci located farther apart on the chromosome will tend to recombine with a probability related to their distance apart on the chromosome. When sets of alleles are closely linked and thus travel together, any phenotype they produce as a result of interaction is more frequent in the population. Increase in recombination rate increases the number of possible phenotypes for the population. Recombination can occur at a *maximum* rate per generation of 0.5 between genes located on the same chromosome, but it always occurs at this rate between genes from different chromosomes. Herein lies the population genetic, or adaptive, value of altered chromosome number, whether it occurs by fusion or breakage. Increase in the number of chromosomes, which arises by breakage and recovery of existing chromosomes, increases overall recombination rates and the possible phenotypic diversity within the population. Decrease in chromosome number via fusion of ancestral chromosomes reduces recombination rate. How this relates to fitness is discussed at the end of this chapter.

Polyploidy is the presence of an extra chromosome or chromosomes compared with the usual number for the species. In organisms in which polyploidy is viable, the extra chromosome(s) can provide material for mutation and increased variability without loss of the original genetic structure.

Inversion is the switching end for end of a segment of chromosome. In inversion heterozygotes normal meiotic events are disrupted; synapsis and chromosome pairing are disturbed prior to the first meiotic division. Inversions suppress recoverable crossovers. If a single crossover occurs within a relatively short inversion, the chromatids that took part in the recombination enter meiosis as either dicentrics or acentrics. The acentrics are lost because they have no centromeres and, therefore, no guidance to daughter nuclei; the dicentrics with two centromeres are caught in a tug-of-war. In both cases inviable meiotic products result. In addition, crossing over is reduced near the breakpoints of inversions. Thus any genetic material located within inverted segments of chromosomes is effectively prevented from recombining. With a recombination probability near zero, inverted segments are tightly linked groups of genes. Their considerable importance in natural populations is discussed later in this chapter (see p. 44).

Translocation is another chromosomal aberration that occurs when two nonhomologous chromosomes break simultaneously and then exchange segments. It is of adaptive importance in some plants. In some species translocations act to effectively link chromosomes together. In most species translocations cause high levels of sterility in translocation heterozygotes.

Migration

In terms of its effect on the allele frequencies of a population, migration is similar to high intensity point mutation. As in the treatment of mutation and allele frequency, we assume that both forms, or alleles, of a gene are present in a single population or in different, partially separated populations. In order for migration to have any noticeable effect on allele frequencies, populations between which some exchange of individuals and genetic information occurs must have different allele frequencies at the loci of interest. Migration is understood statistically as follows: Assume that two populations are interbreeding to the extent that a proportion, m, of the members of population A migrate from population A to population B during each generation; migration is assumed to be at random with respect to the genotype of the migrant. Then at every generation, genotypic migration will occur in accordance with the migration rate and the frequency of the genotype.

The calculations described below may be followed

TABLE 2-4. The effect of interpopulation migration on gene and genotype frequencies

	POPULATION A			POPULATION B		
Genotype	XX	Xx	xx	$X'X'$	$X'x'$	$x'x'$
Frequency	0.09	0.42	0.48	0.64	0.32	0.04

The effects of 5% reciprocal migration:

Genotype	XX'	$Xx', X'x$	xx'
Original frequency	0.09	0.42	0.48
Frequency added from Pop. B	+(0.64)(0.05)	(0.32)(0.05)	(0.04)(0.05)
Frequency lost to Pop. B	−(0.09)(0.05)	(0.42)(0.05)	(0.49)(0.05)
Genotype frequency after 1 generation random migration (rate: 0.05)	0.1175	0.4150	0.4675

in Table 2-4. Initially, the frequency of the allele X is $p^2(XX) + \frac{1}{2}(2pq(Xx))$. But after migration, $m(p^2(XX) + 2pq(Xx) + q^2(xx))$ have left population A and $m'[p'^2(XX) + 2 p'q'(Xx) + q'^2(xx)]$ have joined population A coming from population B. The new frequency of XX after migration is $p^2 + m'p'^2 - mp^2$. Similarly, there are now $2pq + m'2p'q' - m2pq$ heterozygotes and $q^2 + m'q'^2 - mq^2$ xx homozygotes in population A. The new frequencies in population B can be calculated in a similar manner.

Table 2-4 shows that when only 5% of each population actually contributes to the migration, the changes in gene frequency are not really very important per generation. However, migration rates only slightly higher than 5% would suggest to the population geneticist that the populations are actually one interbreeding unit. Even at the rate of 5% interchange per generation, only a few generations of interpopulation migration would be necessary in order to remove any genetic differences between partially separated populations. Additional calculations will show that quite high migration rates and large differences in allele frequency between the intermigrating groups are necessary to suggest deviation of gene frequencies from Hardy-Weinberg equilibrium.

Migration is another population genetic process that appears to have little effect upon Hardy-Weinberg frequencies. What, then, is the importance of interpopulation migration in natural populations? First, as we have just seen, rather high-intensity migration tends to increase the size of the statistically panmictic unit, the genetically important population. The increase in size will prove to be of importance in light of sampling effect (see p. 25). Migration, by introducing variability, can tend to slow rates of loss of genetic variability that result from small population size and associated random drift of allele frequencies.

In addition, interpopulation migration tends to build immediately adaptive variability into participating populations. To see how this works we can examine some data on *Drosophila pseudoobscura,* a type of fruit fly. Prakash and co-workers studied several pop-ulations of *D. pseudoobscura.* Their study covered the area encompassing the American West and Southwest, as well as a population in South America, including Bogotá, Colombia. They found little genetic differentiation at twenty-four loci among these populations, despite the vast distances between them. This result was, of course, rather surprising, since climate, soil types, and food sources must vary over such a large area. Differences in these selective parameters should cause large differences in genetic composition among populations. In order to discover the reason for this lack of genetic differentiation among populations, Lewontin and Prout have continued their studies. They found that flies move from isolated mountain watering spots down temporary streams to the open desert. This happens following spring rains. The flies from many populations probably interbreed in these desert localities. Lewontin and Prout cannot be sure that return to mountain localities and genetic mixture of populations occur, although this seems likely. If they can document once yearly high-intensity migration, what does it mean to the participating populations in terms of genetic variability? Each summer the populations on the mesas experience quite different sets of environmental conditions. Local differences in direction and intensity of selection are expected during the summer and fall, when the flies are active. As a result of these seasonal selection differences within localities, populations might differ during at least part of the year. The local population will partially adapt to local conditions during each summer; but each spring these locally adapted populations apparently intermingle. This results in each of the populations starting the next season of selection with genetic variability from many different local populations, each of which might have been adapted to a quite different set of conditions. Thus each local population is equipped to handle not only the average environmental variations of its own particular area but also those of other areas. Migration in this case both maintains a population's genetic variability and enhances variability and the adaptabilities of the exchanging populations. Such migration might also explain the initial finding of no differentiation among populations of adjacent mesas; the investigators surveyed the genetics of the populations in the spring after the time of intermingling and before the time of summer selection.

One further point might be made about the *D. pseudoobscura* populations. Because of recombination of genomes, completely new genetic types will result; the range of variation of the recombinant types is enhanced by the migration-induced population variation.

Although most of the interpopulation migration Lewontin and associates might find will likely be among adjacent mesas, interpopulation migration might be as important on a wider geographic scale. If each mesa along a geographical transect is identified as a "stepping stone," migration in any direction may be from one to many steps at a time; but the likelihood of multiple steps decreases as the number of steps increases. Thus it is quite likely that a genotype will spread to an adjacent mesa in one rainy season, but multiple-step moves are far less likely. Even so, a phenotype, if favored by selection at each step, might eventually cover much of the range of the species. Since steps can be taken in any direction, but with probabilities of success depending on the fitness of the migrating phenotypes, large-scale interpopulation homogeneity could result. A proper treatment of this concept would need to take rate of breakdown of gene complexes into account and would be very difficult to approach analytically.

So far we have considered only short-term advantages of migration. However, species could adjust to large-scale climatic change by these means. If the environment cools slowly (as it is doing), northern populations of *Drosophila* may be first selected by low temperatures, but they could then contribute genetic information to more southern populations. Consequently, the southern populations would already contain the genetic variation necessary to adapt to cooler conditions by the time the environment cools. Whether or not such prior adaptation is of any advantage to species must be solved analytically.

Population size and genetic structure

Mutation and migration are two processes that affect the genetics of populations. Small population size is a third factor; one of its results is random changes in the genetic composition of a population.

Small population size may result in differentiation of local populations, increased or decreased species variability, and the founding of genetically new subpopulations.

Sampling effect. Whether a small population will benefit as a result of its size and relationship to other small populations or will succumb to a gradual loss of variability and adaptability depends largely on its size and breeding structure. Deviations resulting from sampling (chance deviation from the expected gene frequencies) increase as population size decreases (at least according to simple models of population structure). If, on the other hand, females of a population breed more than once, the age structure of the population determines the speed with which allele frequencies change randomly.

The binomial square law, which is the basis of the study of population genetics, may be restated as follows: *On the average,* the probabilities of two independent events taken two at a time occurring in any combination are given by expansion of $(p + q)^2$, where p and q are the frequencies of alternative alleles at a segregating locus. The important point is that instead of accepting expected genotype frequencies, we now realize the fact of chance deviation of genotype frequencies from those expected.

Predictions of the binomial square hold only for an infinitely large number of trials or infinite population size; the phrase "on the average" in the above statement is of great importance. As the size of the panmictic unit decreases, chance deviations from expectation become more important.

A gambler knows that even if he is playing with an honest coin which is expected to give 50% heads and 50% tails over a large number of tosses, the actual course of his luck may be very different, as shown in

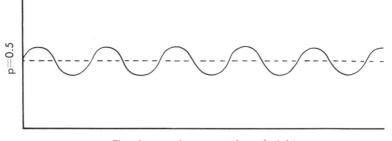

Fig. 2-3. Schematic representation of random fluctuation of heads and tails. This suggests the basis of random drift of allele frequencies in small populations.

Fig. 2-3. Over many thousands of trials, heads do indeed appear half the time, but smaller numbers of trials give rather aberrant results; runs of heads follow runs of tails, and so on.

In a sense, populations flip coins every generation at mating time. In an experiment that you can repeat, coins were flipped two at a time to correspond to the random mating of two gametes, each of which could be either *H* or *T*. In this experiment there were two kinds of populations, one consisting of ten individuals each generation, the other consisting of only five. Ten populations of each size were constructed; results are shown in Table 2-5. Remember that we expect $p = 0.5$, $q = 0.5$, and $(p + q)^2$ to yield 25% *HH,* 50% *HT,* and 25% *TT.*

Two important results emerge from this simple experiment. First, as population size decreases, variation in *genotype frequency* increases greatly. At a population size of ten, few populations differ extensively from expectation; in the smaller populations variation is extreme. The second effect is also important: an *average excess of homozygotes,* or loss of variability is apparent in the smaller populations.

Variations in genotype and allele frequencies. The extent of variation in genotype and allele frequencies depends, in the simplest case, on two factors: (1) population size and (2) allele frequency at the start of a generation. As population size, *N,* and initial product of allele frequencies, *pq,* decrease, expected variation in allele frequencies among progeny populations,

$pq/2N$, increases. This relationship becomes obvious from Table 2-6, where values are plotted after one generation of random mating. Notice that within a fairly wide range of *pq,* variance, or magnitude of chance deviation, of allele frequencies from those expected is essentially constant. These deviations occur with the mating of each succeeding generation.

One result of the sampling effect is *random drift* of allele frequencies. Using the example populations of Table 2-5, we can construct what will likely happen to their allele frequencies after many generations. When analyzing Fig. 2-4, which shows the results of drift, pay particular attention to the bell-shaped curves drawn around the allele frequencies expected for each generation. These represent probabilities of allele frequencies possible at the formation of the next generation.

In Fig. 2-4 population (or line) 1 was initially $p = 0.5$, then $p = 0.8$. These results are from the coin population's first and second generation. At the beginning of the third generation, allele frequency can fall anywhere between 1.0 and 0.0. However, the shape of the curve indicates likelihood of various possible allele frequencies; it is a representation of the *probability* distribution of expected allele frequency. The expected value is most probable; values differing from this become less probable as difference increases. Therefore, at any generation, allele frequency can be nearer 0.5, or nearer zero, or 1; but once gene frequency is substantially different from 0.5, likelihood of return to this value is small.

TABLE 2-5. Variation in genotype frequencies among synthetic populations of size 5 (a) and 10 (b)

GENOTYPE	POPULATION (SIZE a)										TOTALS (AVERAGE FREQUENCIES)
	1	2	3	4	5	6	7	8	9	10	
HH	0.4	0.4	0.2	0.2	0.0	0.2	0.4	0.6	0.2	0.4	0.30
HT	0.4	0.4	0.4	0.6	0.8	0	0.4	0.2	0.6	0.4	0.42
TT	0.2	0.2	0.4	0.2	0.2	0.8	0.2	0.2	0.2	0.2	0.28

GENOTYPE	POPULATION (SIZE b)										
	1	2	3	4	5	6	7	8	9	10	
HH	0.2	0.3	0.3	0.3	0.4	0.1	0.3	0.2	0.2	0.4	0.27
HT	0.3	0.2	0.5	0.6	0.6	0.8	0.5	0.6	0.6	0.1	0.48
TT	0.5	0.5	0.2	0.1	0.0	0.1	0.2	0.2	0.2	0.5	0.25

TABLE 2-6. The relationship between initial allele frequency and expected variance of change in allele frequency of progeny as a function of population size

2N	p 0.5	0.4	0.3	0.2	0.1
10	0.025	0.024	0.021	0.016	0.009
20	0.0125	0.012	0.0105	0.008	0.0045
50	0.005	0.0048	0.0042	0.0032	0.0018
100	0.0025	0.0024	0.0021	0.0016	0.0009

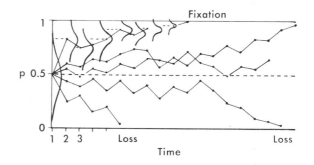

Fig. 2-4. Changes in gene frequency because of genetic drift, which can occur in any population. Variance of expected change between progeny and parental allele frequencies is shown by the bell-shaped curves and defined as $pq/2N$. Variance increases with decreased population size but decreases with deviation from allele frequency equality ($p = q = 0.5$).

Loss of variability within a population. The eventual result of drift is the loss of one or the other allele at a drifting locus—the line is "fixed" for the remaining allele. Of course, it is now a genetically invariant, monomorphic line with respect to the fixed locus. Our other imaginary lines in Fig. 2-4 have been drifting, too, and some have become fixed.

Fig. 2-5 shows the process of drift in laboratory populations of *Drosophila melanogaster,* the common fruit fly. The 105 lines in the study were maintained at a population size of 16, and initially the brown 75, (b^{75}) allele for eye color was present at a frequency of 0.50

in each population. Drift is obvious from Fig. 2-5. Lines immediately begin to deviate from 0.50 b^{75}; and by the fourteenth generation many populations are either fixed for or have completely lost the allele. There is much variation in frequency among populations.

One result of drift is loss of variability *within* a population; another is increased variation *among*

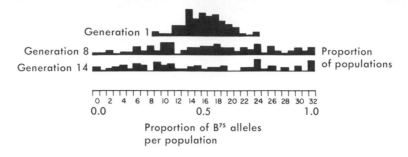

Fig. 2-5. Sampling effect. Data are of 120 populations of *Drosophila*. Each population contained 16 parents at the beginning of a generation. The initial lines all contained 16 brown[75] and 16 brown alleles for eye color. Already by generation 1 considerable variability has developed among lines. Fixation of one allele is evident by generation 8. Loss of genetic variability is quite rapid. (After Buri, P. 1956. Gene frequency in small population of mutant *Drosophila*. Evolution **10**:367-402.)

populations. Both effects are important to the genetics of natural populations.

Fixation, or the loss of variability at a locus, usually does not occur with respect to one gene in all of a large number of populations. However, a small population often becomes homozygous by chance for a large number of different loci. Each loss of variation resulting from homozygosity reduces the genomic variability of the population and, ultimately, the population's adaptability is reduced.

Some geneticists have suggested the homozygosity resulting from small population size is part of the reason endangered species, such as eagles, pelicans, and whooping cranes, may face eventual extinction. Most of these populations are small because of range reductions or predation by man, but the resulting small population size should contribute to their eventual demise, because they, like all species, live in fluctuating environments that require constant adaptation. Small population size reduces genetic variability and hence the ability to adjust.

Variability among populations (adaptability). Sewall Wright argued many years ago that random processes can contribute to adaptability of species. Some plants and many insects have very high rates of increase (see p. 51), and many of these are colonizing species that move to new environments at each genera-

tion. It is possible to develop the thesis that all natural species that live in fluctuating environments are colonizers and that many of them depend for their existence on their ability to adapt. Theoretically, these species often contribute small samples of the parental population to the next generation's environment, or to a diversity of environments unlike those faced by the parental population. Each propagating population is a random sample of the genomes carried by the parental group, and of recombinant genomes not present in the parental population.

Some such colonies fail to survive because they encounter environments for which they are completely unsuited. Others, however, become established in favorable habitat, which may differ from the environment of the parental population. The long-term result of such a process is shown schematically in Fig. 2-6. The heavy bell-shaped curve represents the range of environmental tolerance of a parental or ancestral population. Each subsidiary bell is that of a colony. Notice that (1) these colonies differ in environmental tolerance from the parental population and (2) more importantly, they in total represent a widely adapted species.

The above statement assumes that some migration and genetic intermingling occur among the populations. The evolutionary result of such strategy should

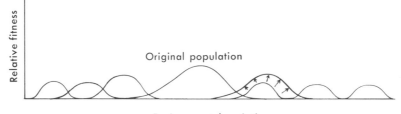

Fig. 2-6. Possible consequences of sampling effect. Subpopulations are formed, each consisting initially of a random sample of parental population genomes and phenotypes. Selection of these "founder populations" can result in a complex of phenotypically different subpopulations. Arrows indicate that any founder population may broaden its original phenotypic distribution, assuming interpopulation migration and recombination.

be that the species, via its separate populations, is adapted to a wide range of possible environments and is unlikely to succumb to an abrupt change in some environmental parameter.* The first stages of this process are called the "founder effect," which refers to the propagation from founders of new colonies that are different genetically from ancestral populations.

Sampling effect and evolutionary potential. A population does not have to be continually small for sampling effect to be important to its evolutionary history and potential. It has been found that if a population fluctuates in density or size, as most do (some violently), the smallest size that the population reaches is the most important determinant of its effective size.

Some populations periodically pass through genetic bottlenecks; their size at times becomes small enough that random drift of gene frequencies is important. If this bottleneck is combined with some selection, the evolutionary possibilities of the population may be enhanced through chance occurrence of adaptively advantageous gene arrangements in high enough frequencies for selection to be effective. The speed of evolution, as we shall see in the section on natural

selection, is partially a function of the frequency of the genetic type being selected.

The importance and process of sampling are obviously closely associated with recombination. Genetic linkage is involved, but it and the other factors already discussed that tend to control the genetic variability of populations can be better appreciated with a full knowledge of natural selection, a nonrandom process.

NATURAL SELECTION

In nature, all aspects of the environment, such as competition, predation, parasitism, weather, and the ionic concentration of soil and water, exert continual selective force on a population. There is even selection within a population because of the presence or absence of other genotypes. Because of sampling effect, small population size is of selective significance to the population.

The working of selection depends on the presence of genetic variability within the population. An intuitive example of how selection works relative to one environmental parameter has already been presented. After the concept of ecological niche has been developed, it will be possible to investigate the effects of selection on many parameters at once. At present it will be better to think of selection relative to only one genetic locus and one, or at the most two, environmental variables at a time.

*Levins (1965) notes that such populations have maximal ability to adapt to long-term changes in their environment, whereas their genetic reaction to short-term environmental "noise" is reduced.

Selection acts to cause changes in allele frequency and is an expression of the fact that some genotypes are less able than others to survive in environments of the moment. These genotypes are said to be less "fit" than others, meaning that they are less able to contribute to the next generation than are others. *Fitness,* then, refers to the relative ability of members of a genotype to survive to reproductive age and to produce offspring as compared with members of other genotypes or with the total population.

By convention, the scale of fitness of genotypes is usually established relative to a standard value of 1. The genotype that is most fit has a fitness value of 1.0, and other fitnesses are adjusted accordingly. Some confusion has arisen from this convention; this scale of relative fitness is quite different from a population's rate of increase. Reduction in relative fitness need not imply a reduced rate of increase for a population.

Selection intensity

Selection intensity is reduction in fitness (from 1) of a genotype. It refers to the relative inability of a genotype to survive and reproduce, and it can mean either that a certain excess proportion of the genotype dies before reproduction, or that members of a genotype produce fewer gametes than those of other genotypes; or it can mean some combination of these factors. (As shown in Chapter 3, fitness and selection intensity are related to length of the reproductive span of members of a genotype in relation to that of the rest of the population.) If 10% of the members of a genotype fail to survive, selection intensity against that genotype is 0.1, and its fitness is 0.9, assuming that all genotypes produce the same number of gametes once reproductive age has been reached.

Effects of selection on a population

It is of more than academic interest to be able to calculate the effects of selection on a population. To illustrate, the following convention is established. Three genotypes in a population are:

$$AA \qquad Aa \qquad aa$$

Their fitness is represented as:

$$w_{11} \qquad w_{12} \qquad w_{22}$$

Their selection intensities are as follows:

$$1 - w_{11} = s \qquad 1 - w_{12} = t \qquad 1 - w_{22} = r$$

The following is a completely intuitive method of calculation. We will then develop a method of predicting the short-term course of evolution of a population under selection of any form. Assume that the initial genotype frequencies for eye color in *Drosophila melanogaster* are as follows:

Genotype	BB	BB^{75}	$B^{75}B^{75}$
Phenotype	Brown	Orange	Light orange
Frequency	0.25	0.50	0.25

Also assume that 10% of the brown-eyed flies fail to reach reproductive age. By convention, selection intensities for the three genotypes are $s = 0.1$, $t = 0$, and $r = 0$; relative fitnesses are 0.9, 1, and 1, respectively.

To calculate the frequencies of the genotypes after one generation of selection has passed, it is logical to determine what the relative contributions of the genotypes are to allele frequencies in the population's gamete pool or at mating time. To estimate the frequencies at reproduction, simply multiply the frequencies of the genotypes by the proportions that survive (their fitnesses):

Genotype	BB	BB^{75}	$B^{75}B^{75}$	
Frequency	0.25	0.50	0.25	
Fitness	0.9	1.00	1.0	
Product	0.225	0.50	0.25	Sum: 0.975

However, the sum of these values is only 0.975 and, therefore, these are not frequencies. We need to adjust by the total fitness of the population (total proportion to survive) to obtain genotype frequencies:

BB	BB^{75}
0.225/0.975 = 0.2308	0.5/0.975 = 0.5128

$$B^{75}B^{75}$$
$$0.25/0.975 = 0.2564$$

It becomes a simple matter to calculate the composition of the gametes of these alleles and then to calculate genotype frequencies at the beginning of the next generation. We use the tool we already have, $p = p^2 + \frac{1}{2}(2pq)$; $q = 1 - p$. Then:

$$p^2 \ (BB) = 0.2374$$
$$2pq \ (BB)^{75} = 0.4996$$
$$q^2 \ (B^{75}B^{75}) = 0.2630$$

This method is intuitive, computationally messy, and only structurally informative. It does, however, suggest the logic behind a more formal and useful method of calculating the effects of selection. First, we want to know the frequency of one allele at the time of reproduction of the population. If there were no selection, this would be $p' = p^2 + \frac{1}{2}(2pq)$. With selection against BB, we have $p' = \dfrac{p^2(w_{11}) + pq}{1 - p^2(s)}$ because only a proportion, w_{11}, of the AA homozygotes survive to reproduce. The denominator of the equation takes into account that total population fitness is reduced by the proportion of the population $(p^2 s)$ that dies before reproduction. Thus the denominator is the population's mean fitness. The change, Δp, in gene frequency per generation is:

$$\Delta p = p' - p = \frac{p^2 w_{11} + pq w_{12}}{1 - p^2 s} - p$$

In Table 2-7 this expression is made more manageable algebraically and is listed along with several other expressions of Δq that are of interest. These expressions are useful to predict the direction and speed of change of gene frequency in populations subject to natural selection.

Analysis of the above equation leads to an interesting observation. Table 2-7 shows that speed of gene frequency change is determined by initial frequency and by selection intensity. Moreover, in this particular case speed of change is related, because of allele frequency, to the proportion of the allele that is in the

TABLE 2-7. Formulae used to predict change in gene frequency per generation

Genotype	A_1A_1	A_1A_2	A_2A_2
Initial frequency	p^2	$2pq$	q^2
Selection intensity	s	r	t
Fitness	$1 - s$	$1 - r$	$1 - t$

If selection is against A_1A_1:

$$\Delta p = \frac{p^2(1 - s) - pq}{1 - p^2 s} - p$$

A_1A_1 and A_1A_2:

$$\Delta p' = \frac{p^2(1 - s) + pq(1 - r)}{1 - p^2 s - 2pqr} - p$$

A_1A_2:

$$\Delta p' = \frac{p^2 + pq(1 - r)}{1 - 2pqr} - p$$

A_1A_1 and A_2A_2:

$$\Delta p = \frac{p^2(1 - s) + pq}{1 - p^2 s - q^2 t} - p$$

A_2A_2:

$$\Delta p = \frac{p^2 + pq}{1 - q^2 r} - p$$

homozygous condition (and selection-prone) relative to that proportion in the heterozygous state.

Two conclusions emerge from this case of selection against a homozygote. First, speed of change of allele frequency is greatest for any particular selection intensity when allele frequency is near 0.5, and it decreases as the frequency of the allele selected against decreases (Fig. 2-7). Second, in an infinitely large population the allele selected against will never be lost. Table 2-8 shows the reason for this: As allele frequency decreases, greater and greater proportions of the allele are in the heterozygous, protected form. When selection is against both a homozygote and a heterozygote it is most effective at initial frequencies of around 0.6 to 0.7.

The above observation is of some importance, if only because variability of the locus is maintained in large populations. In addition, allele frequency change, Δp, slows down as the selection begins to cost the population less. The cost of selection per generation is measured by the population's mean

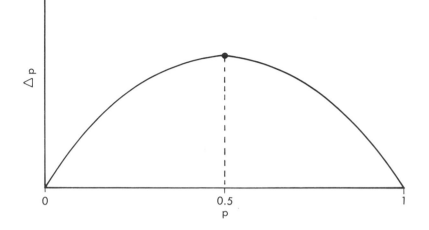

Fig. 2-7. Relationship between allele frequency before selection and expected change in allele frequency. All points on this curve assume the same set of selection intensities. Actual values for Δp depend on actual fitness values. This figure only applies to a gene with no dominance. The relationship between allele frequency and effectiveness of selection is not exact (results depend on whether selection favors or disfavors the allele).

TABLE 2-8. Population composition relative to a disfavored allele when selection is against the homozygote only*

p (FREQUENCY IN POPULATION)	2pq	p^2
0.5	0.50	0.25
0.4	0.48	0.16
0.3	0.42	0.09
0.2	0.32	0.04
0.1	0.18	0.01
0.05	0.095	0.0025
0.01	0.0198	0.0011
0.001	0.001998	0.000001

*The result of this pattern is that, theoretically, the allele frequency can only decrease by one-half each generation. Practically, sampling effect will cause loss of the disfavored allele in all but infinitely large populations.

fitness. In this manner time becomes available for recombination and perhaps amelioration of selection intensity.

Directional selection

Directional selection against a dominant allele results in eventual loss of that allele, but remember that the biochemical expression (as regards dominance) of an allele can change depending on its genetic company.

A well-documented case of directional selection is that resulting from an Australian attempt at biological control of rabbits. This case is also an example of coevolution where two interacting species evolve to their mutual benefit. The coevolution idea is useful for future reference because it indicates processes involved in integration of community structure.

Rabbits were introduced to Australia in 1859 and rapidly multiplied until, by 1950, they constituted a serious economic problem because they competed with sheep for grass. Myxoma virus, a fatal parasite of rabbits, was introduced into the rabbit population in 1950, and within 1 year 99.5% of the rabbits had died. By 1952, however, the effectiveness of the virus had apparently decreased, and the rabbit population was nearly as large as it had been before the introduction of the virus.

Investigation revealed several reasons for this. Both the rabbit population and the virus population had reacted to increase their fitnesses. Myxoma virus, like many other parasitic organisms, is completely dependent on its host for survival. Myxoma virus is vectored (transferred) from host to host by mosquitoes that attack only live rabbits. Thus the birth rate, or reproductive fitness, of the virus is a direct function of the life span of its host rabbit, because longer host life results in greater chance that subsequent transfer of virus will occur. Natural selection should favor less virulent virus particles, which allow their host to live a normal life span.

Viruses have only one chromosome, and the Myxoma virus introduced in 1950 was genetically monomorphic. Therefore, genetic variation necessary for evolution must have arisen via chance virulence-reducing mutations. Since evolution of the virus had certainly occurred, investigators sampled the Australian rabbit population in 1952 and found not only the original strain of virus, but also four other strains, which were apparently the result of fortuitous single gene mutations in the virus population. These strains and estimates of their respective virulence are listed in Table 2-9. Compare the strain frequencies in 1952 with those of 1958. It is clear that very rapid unidirectional selection for low virulence occurred among the viruses and that the raw material for selection was a series of mutants, each of which acted to reduce the virus's virulence.

Just as the virus population evolved reduced virulence, which amounted to an increase in fitness, the rabbits evolved resistance to viral infection. In 1950, virus strain I was frozen and saved for future tests. In later years wild rabbits were tested with this virus. The results are shown in Table 2-9. Resistance of the rabbits has increased. The rabbits and the virus evolved in such a way as to coexist, if not to mutual benefit, at least to mutual reduction of detriment. Coevolution, as stated previously, is the rule in natural communities; numerous other examples are well documented. In one example involving leaf-cutting ants and their fungus crop, the ants keep their fungus well cultivated and reap a return in foodstuffs. The ants control the growth form of the fungus.

Balancing selection

Other types of selection exist. Most can be grouped under the broad generic term *balancing selection*. These types are an important mechanism for the maintenance of variability within a population. Basically, balancing selection depends on the heterozygote at a locus having a higher fitness than either homozygote. Such sets of fitnesses can occur in many ways; some of these form the basis of much ecological theory, which will be detailed shortly.

If selection unconditionally favors heterozygotes, algebraic manipulation of the proper equation of Δp (see Table 2-7) shows that each population has a characteristic stable (equilibrium) gene frequency, $\hat{p}$, which is maintained once it is reached. As

TABLE 2-9. Coevolution of rabbits and myxoma virus

	VIRUS STRAIN				
	I	II	III	IV	V
$\bar{x}$ days rabbits survive	13	14-16	17-28	29-50	Normal
Mortality rate (%)	99.5	99	90	60	30
1958 frequency of strains	0	0.25	0.56	0.14	0.05
Percent death caused by strain I virus In:					
1953	95				
1954	93				
1955	61				
1958	54				

$\hat{p} = \dfrac{w_{12} - w_{22}}{2w_{12} - w_{11} - w_{22}}$ shows, the value of this equilibrium gene frequency is solely dependent on the relative fitnesses of the genotypes at a locus. As shown in Fig. 2-8, approach to this frequency is asymptotic; as the equilibrium is approached, selection becomes more balanced and has proportionately less effect on allele frequency.

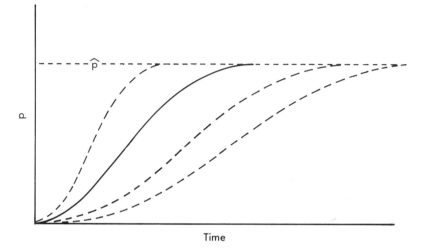

Fig. 2-8. Course of approach to equilibrium allele frequency. Selection intensity increases from right to left. Faster maximum rate of change of allele frequency results.

TABLE 2-10. Hospital data pertaining to sickle cell anemia in Africa

	TYPE OF ADMISSION	
	AS (HETEROZYGOUS FOR SICKLE CELL)	*AA* (NORMAL HEMOGLOBIN)
Simple malaria (not fatal)	13	70
Cerebral malaria	0	47
Black water fever	0	6
Total deaths	0	53

Total admissions for all causes: 688
Selection against normal hemoglobin: 53/688 = 0.09

$w_{AA} = 0.91$
$w_{AS} = 1$
$w_{SS} = 0$

$$p = \frac{(1 - r) - (1 - t)}{2(1 - r) - (1 - t) - (1 - s)} = \frac{1 - 0.91}{2 - 0 - 0.91} = \frac{0.09}{1.09} = 0.082$$

Many natural polymorphisms are the result of this form of selection. A famous example is that of the allele for sickle cell anemia, Hb^s, in man. When homozygous, this allele results in a flaw in the structure of the hemoglobin molecule, which reduces the oxygen-carrying capacity of the blood. One would suppose that such a disease, which is fatal only as homozygous $Hb^s Hb^s$, would have a very low frequency in a population. However, in Africa where this sickle cell disease is prevalent, malaria is an important cause of mortality. Sickle cell *heterozygotes* are largely immune to both the malarial and other parasitic infestation. In addition, for some unknown reason, heterozygous females are likely to have more children than normal females. Because of these factors, sickle cell heterozygotes have higher fitness than either homozygote, and the allele is maintained by the balance of selective forces.

In one section of Africa the frequency of the Hb^s allele is 0.11. Using selection intensities shown in Table 2-10 which do not include all selective forces important to this locus, we predict an equilibrium frequency of 0.082.

In the United States, malaria has not been a problem since about 1900, and since then selection against sickle cell anemia has been directional. Assuming a generation interval of 20 years (and thus nearly four generations of selection since 1900), we predict a present frequency of 0.1. The initial frequency (0.20) corresponds to an average frequency for the Hb^s allele among blacks on the African mainland during the slave trade. The *predicted* United States frequency among blacks is 0.1 and is calculated by assuming that selection occurs only against Hb^s homozygotes in the absence of the balancing effect of malaria. In Alachua County, Florida, the frequency is 0.1 among blacks, so it appears that we have correctly analyzed the situation.

Although other cases of this type of heterotic selection could be cited, most polymorphism is the result of balancing ecological factors.

Balancing ecological factors and adaptive strategies. Basically, balancing ecological factors

occur whenever selection intensities fluctuate over short periods of time (for example, from generation to generation) or when a population occupies two or more different environments in which selection intensities favor alternative alleles. The theory of this form of selection and related evolutionary strategies was developed by Richard Levins and others.

Levins suggests that natural populations live in either of two kinds of environments, coarse- or fine-grained. Coarse-grained environments are those in which (1) large fluctuations occur with a periodicity that is long is comparison to the generation time of the population being considered or (2) the environment is divisible spatially into patches easily distinguished by members of the population. The population is expected to consist of genetic types able to live in either of the alternative forms of the spatially coarse-grained environment. A tacit assumption of Levins' analysis (and of all population biological theory) is that the mean fitness, $\overline{w}$, of the population should be maximized.

Some introductory groundwork is necessary before we approach the study of Levins' theory of adaptive strategies. Remember that there are at least two ways an organism or population can handle environmental variation. Physiological homeostasis is one possibility; members of the population may be able to acclimate to fluctuations in environment. Developmental homeostasis is another possibility; members of some species are able to develop in ways determined by their environment. An organism might grow to large body size in cold environments and small size in warmer environments. Plants grow large leaves in areas lacking abundant sunlight and small ones in sunny locations.

Quantitative traits and natural selection. Most phenotypes are controlled by not one, but a complex, of genes. Such traits as stature, weight, bill size and shape in birds, feeding habits, and other ecologically important behavior pattern vary quantitatively. Single genetic loci control only a few characteristics. Because of the large number of loci that control them, quantitative traits approach a normal distribution of phenotypes; for each trait there is a mean or most common, expression, and bounding it there is a series of other phenotypes that occur less frequently in a population. Since such traits are controlled by many, sometimes independent loci, the way in which component phenotypes are selected by ecological variables depends on how many loci are involved and on the extent to which each locus exerts influence on phenotype. As the number of loci influencing a trait increases, the average importance of each decreases. Of course, these loci can interact in various ways.

Selection can operate only to change allele frequencies, gene arrangements, and other aspects of genetic control; it is not operative on environmentally induced variation. This is a two-edged sword. On the one hand, a trait with a great amount of environmental variance is expected to be stable in the face of small environmental change; physiological modifications act to change phenotype within the lifetime of a single individual and selective death is avoided. Thus populations are buffered against essentially useless genetic change due to environmental "noise." On the other hand, this same genetic homeostasis and physiological conformability of individuals implies that genetic change under longer and more selectively significant environmental change will be slower than if phenotypes were more fully controlled by genetic factors. Because we cannot visualize with any confidence the gene interactions that contribute to quantitative traits, the theory of selection for them is based on expectations of selection on a normal distribution. No attempt is made to analyze or predict changes that may occur at specific component loci. We can derive mathematical expressions to predict the rate of change of overall genetic variance.

Threshold selection. Under this form of selection, we assume that all phenotypes beyond a threshold point are killed or fail to reproduce and that the remainder of the population contributes to the next generation (Fig. 2-9). By using statistics appropriate to a normal distribution, investigators have determined that change in the mean of the trait is dependent (1) on the selection differential between the before and after selection populations and (2) on the proportions of the variance

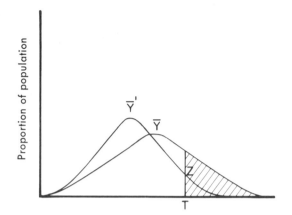

Fig. 2-9. Threshold selection of a quantitative trait. *T* is the selection threshold; phenotypes in the shaded area do not reproduce. $\overline{Y}$ and $\overline{Y}'$ are respectively pre- and post-selection mean phenotypes. For a mathematical derivation see Crow and Kimura, 1970.

that are genetic and environmental. Progress of selection is affected by the ratio between total variance (genetic and environmental) and genetic variance. Thus traits that are heavily affected by environmental modification of phenotype respond slowly to selection and are more stable.

By defining the threshold point, we assumed only that members of a population with a particular phenotype would fail to reproduce, but assigned no cause. If the phenotypes were controlled primarily by genetic differences, reproductive failure would be of selective importance, a loss of Wrightian fitness. A change in mean phenotype would occur in the progeny generation. If, on the other hand, most population variance is environmental (phenotypes change because of physiological or behavioral mechanisms within the lifetime of single individuals) then little genetic death will occur. Little or no difference is expected between mean phenotypes of parental and progeny generations. The progeny simply must adjust to environmental variation just as their parents did.

The degree to which organisms use physiological and developmental strategies to adapt to environmental fluctuation is dependent on many factors, not the least

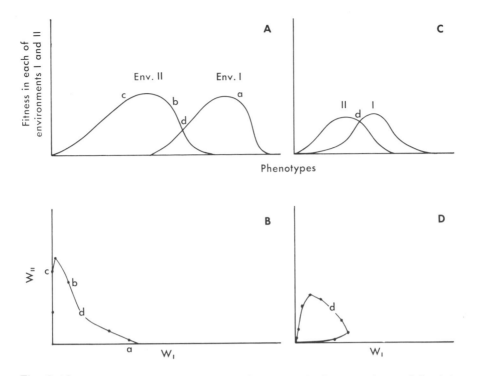

Fig. 2-10. Construction of fitness sets. **A** and **C** represent the fitnesses of a population in two different environments (I and II). Each point on each line represents a phenotype. Thus phenotype **D** has some ability to survive and reproduce in both environments, but phenotype **A** is very well suited for environment I but dies in environment II. In **B** and **D** the fitness of all possible phenotypes are plotted in both environments. Each phenotype is graphed as W_{II} (y axis) − W_I (x axis).

of which is their physiological capacity to acclimate. There are two basic types of organisms, those with great powers of acclimation and low adaptive abilities and others, like some insects, with great adaptive powers. We can analyze the apparent dichotomy in terms of Levins' "fitness sets" and adaptive function.

Levins imagines that the phenotypic tolerance characteristic of a member of a particular genotype can be represented much as in Fig. 2-10. There is an optimal phenotype for each environment in which the species is placed. Redrawn in Levins' terms, Fig. 2-2 looks like Fig. 2-10, *A*. The fitness of phenotypes are plotted for two environments, I and II. If the diagram is drawn as in Fig. 2-10, *B*, we see the plot of fitnesses of all phenotypes in both environments. For example, phenotype *c* is highly fit in environment I, but it cannot survive in environment II.

Any other point on this fitness set can be analyzed similarly. Phenotype *d*, a "jack of all trades," has the same fitness in both environments. If environments are sufficiently similar, there will be some intermediate phenotype composition of the population for which the total fitness of the population is maximum. The fitness set is said to be convex (Fig. 2-10, *C* and *D*), and since the environmental range is less than the tolerance of individuals, the population will be monomorphic regardless of the temporal or spatial arrangements of the environment. It might be ideally monomorphic for some heterozygous genotype, but segregation will result in less fit homozygotes being continually formed.

If, however, environments I and II are dissimilar, the fitness set will be concave (Fig. 2-10, *B*). No phenotype in the population will be able to cope with all of the variation of the environment; some will be best adapted in environment I, others in II. In this situation the genetic and adaptive strategy of the population will depend completely on the graininess of the environment. Whether the population is monomorphic and able to live in only one environment or is polymorphic and able to live in both or all environments will depend on the ease with which the altenatives can be distinguished and perhaps chosen.

In a fine-grained situation, habitat types or alternative environments are in close apposition temporally, and the organism is faced with a succession of different environments. Since it is assumed that the succession occurs during the lifetime of an individual animal during a single generation, there is no opportunity for genetic change. Population fitness is maximized if the population is monomorphic for the more common environmental type. Mammals that acclimate to temperature changes are adapted to fine-grained environments.

Populations face coarse-grained environments as alternatives. For example, one generation or series of generations might be faced with cold, the next series with warm environments. If the temporal variation is too great to be handled by physiological or developmental plasticity or by avoidance mechanisms such as torpor, the population must be genetically polymorphic in order to survive. Such populations routinely adapt to variation in their environment by genetic change.

It would be naive to suggest that environments are either fine- or coarse-grained and that species can be so divided and characterized. Some work done by several investigators indicates problems of definition quite well. The mud snail *Nassarius obsoletus* lives on mud flats that are periodically exposed to air, heat, and fresh water. Since the species also has a wide geographic range one might expect it to exhibit evidence of strong adaptability. In fact, this snail is highly homozygous, possessing little genetic adaptation ability. However, this species can completely change its feeding organs depending on whether its prey of the moment is animal or plant material. Therefore, it is not illogical to suggest that the species is exceedingly good at physiological homeostasis.

In contrast, the relatively low intertidal mussel *Modiolus demissus* shows striking changes in gene frequency over a relatively small geographic range. Furthermore, it and an ecologically similar species, *Mytilus edulis,* have remarkably similar genetic structures and dynamics. Koehn and co-workers, who investigated the relationship between these species, concluded that they are genetically similar and react similarly to environmental fluctuation as a result of long-standing ecological similarity. This is evidence that if species are not too different genetically they should react to environmental modification in similar ways.

Direct evidence of the adaptive value of single gene polymorphism is difficult to find, since most genes must be expressed as functions of an individual's entire phenotype and, therefore, should have only small unique adaptive effects. The allele responsible for the hemoglobin variant expressed as sickle cell anemia and some other genes in the human population are exceptions. A study by Merritt shows a good correlation between the temperature kinetics of allele enzymes of lactate dehydrogenase (LDH) and their frequencies in natural populations of fish. Merritt's work on enzyme kinetics shows that $A'A'$ gene-enzyme has a lower affinity for substrate at high temperature and a higher affinity for substrate at low temperatures than the AA homozygote gene-enzyme. Therefore, we expect that population polymorphism at the LDH locus will contain high frequencies of the A' allele-enzyme in warm regions and low frequencies in colder climates. Merrit's results on allele frequencies of natural populations support this hypothesis very well. In Fig. 2-11 frequencies of the A' allele-enzyme are plotted against July temperature. The decline in A' frequency cor-

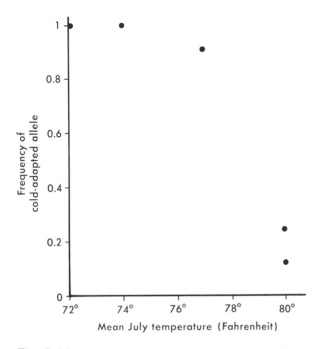

Fig. 2-11. Mean July temperature (degrees F). (Data from Merrit, 1972, and World Almanac.)

responds to the range of mean summer temperatures of 23.9° C (northern sites) to 26.7° C (southern sites). Other studies show similar relations between enzyme function and geographical distribution of allele frequencies.

From consideration of adaptive response at the rather abstract level of single electrophoretic loci we can return to morphological variation. The limpet *Acmaea testudinalis scutum* is found at almost all intertidal levels, meaning that different segments of the population are routinely exposed to quite different selective regimes. Animals that live on rocks far up the beach are uncovered for long periods and are subject to drying and to intense heating and cooling. Those living on lower rocks are almost never exposed to the air for long, but they are subject to pounding by surf and abrasion from sand and shell particles which are propelled about by heavy wave action. A population near

Marineland, Florida, differs within itself strikingly in apparent response to selective regimes to which portions of the population are exposed. Limpets from low rocks are very flat whereas those from high rocks have tall, peaked shells. Those animals found part way up the beach have shell heights intermediate between the two extremes.

Investigation revealed that the population of these rocks is initially homogeneous; young animals from high rocks cannot be differentiated from those originating from low rocks. However, as individuals grow they begin to assume their particular adult shape. High, peaked shells are hedges against desiccation, since these animals have a realtively small surface-to-volume ratio. (Animals with small surface-to-volume ratios should have more tissue water and should be able to withstand drying for long periods.) Low intertidal limpets are squat in response to wave wash and abrasion. Flat shells offer less resistance to the waves and there is less surface for waves to wash upon. Limpets with flat shells ought to be more able to remain fixed to the substrate than animals with more peaked shells.

It is difficult to tell whether such changes in the population are due to selective mortality or to differential growth. Available evidence suggests that most of the changes in shell form are the result of differential growth rather than mortality. No young snail has a shell as peaked as those found among large, high intertidal animals. This is an example of an environment that at first appears to be coarse grained (and indeed is for most intertidal species), but which, at least for one species, is quite fine-grained. If this species did not possess the remarkable developmental plasticity it seems to have, the amount of genetic death involved in colonization of the entire intertidal range at each generation would be immense.

Disruptive selection

Acmaea digitalis illustrates a form of selection that has been called disruptive or centrifugal. Homozygotes are optimal or most fit, and heterozygotes have lower fitness. This situation is genetically less stable than balanced polymorphism where a stable gene frequency is attained; here the polymorphism is maintained by environmental heterogeneity. If the heterogeneity is spatial, further adaptation on the part of the population might involve the animals actively choosing their optimal patch or grain types. If phenotypes are continuous rather than discrete, such behavior serves to maximize population mean fitness by reducing mortality of intermediates.

The limpet *A. digitalis* occurs in a continuum of shell patterns ranging from pure white to dark tan with heavy brown stripes (Fig. 2-12). These forms can be found in a single panmictic population. The environment of the limpet consists of two distinct habitats: white, lightly striped, goose-neck barnacles, and tan

sandstone rock patched with algae. The barnacle colonies are attached to sandstone. Shore birds selectively prey on dark tan limpets living on the white barnacles, as opposed to those on the surrounding tan rock face. Light colored limpets choose to occupy barnacle colonies, while dark tan animals live by choice only on the rock face. Animals of intermediate shell pattern are ambivalent; they exhibit little choice of habitat.

This situation can be represented as a concave fitness set, in which the species occupies a coarse-grained environment. Selection is most intense against intermediates. It is easy to imagine that these are genetic heterozygotes, although no analytical genetics has been done to justify or disprove this assumption. The color polymorphism, together with the behavioral

Fig. 2-12. *A. digitalis* has a variety of shell patterns ranging from pure white to dark tan.

adjustments described above, serve to maximize the fitness of the population.

This disruptive type of selection, if it occurred together with sexual isolation of forms, would result in the formation of two species. It is easy to see that fitness for this species would be low indeed were it not for the genetically determined ability and propensity of differently patterned limpets to choose appropriate backgrounds. Selection without this propensity would be constant (the environment would be fine grained because the limpets would move at random between the rock and barnacles), and a monomorphic population would eventually result. The limpet forms that would persist would be those corresponding to the more common environmental type.

The population probably retains its polymorphic form, instead of speciating, for energetic reasons. Light, barnacle-dwelling limpets must feed on the rock face, and since most limpets occur low in the intertidal zone where selective predation is low and energy content is high, predator selection would be counterbalanced by energetic demands. Light limpets are selected for reduced habitat specificity in the lower intertidal zone.

The presence of most of the population in the low intertidal area also explains why speciation does not occur in these animals. In the higher areas, dark limpets breed earlier in the spring and later in the fall than do the barnacle-associated forms. This might be an initial stage of speciation. In the lower regions, breeding cycles do not differ, and the population remains panmictic. More limpets occur in the lower intertidal area, and gamete production is probably greater in this area of abundant food; therefore, most of the population interbreeds, thereby swamping any incipient speciation.

Frequency-dependent selection

The idea of the predation-maintained concave fitness set may be extended to include many prey species that occur in alternative forms. Sheppard, in a now-famous study, analyzed the cause of a shell color polymorphism in *Cepaea nemoralis,* a European land snail (Fig. 2-13). They found that birds preyed on snails having the most common shell color and pattern. Such selection is called apostatic, or frequency-dependent, and results in maintenance of numerous patterns or morphs in fluctuating frequencies. Thus the snails live in a temporally coarse-grained environment.

Environment for this snail is spatially fine-grained

Bandless   Two varieties of banded

Fig. 2-13. Shell patterns in the European woodland snail *(Cepaea nemoralis)*. This common snail comes in a variety of color patterns, only three of which are shown. (From Dillon, L. S. 1973. Evolution, concepts and consequences. The C. V. Mosby Co., St. Louis.)

since there are many possible habitats for the snails, such as green grass, dead leaves in woods, etc. For each of these habitats, some series of shell colors and patterns will be protectively colored, and therefore more fit. But the snails occur randomly among habitats. There is a final degree of complexity: The environment is temporally coarse grained on a seasonal basis. In fall and winter the grass and woods floor are brown and light colored snails are more frequently preyed upon; in spring and summer the grass and woods floor are greenish yellow, and selection direction reverses with darker snails having the lower fitness.

Frequency-dependent selection occurs in response to factors other than predation. Kojima has found that there are optimal gene-enzyme frequencies in laboratory populations of *Drosophila melanogaster*. Any deviation from these frequencies elicits selection and a return to optimal frequencies. Such selection is probably related to subdivision of the medium in the culture bottles and is really intraspecific competition (see Chapter 4).

Group selection

One further controversial form of natural selection may be essential to phenomena observed in nature. This is group selection. Thus far we have assumed that selection acts on individuals and not on more inclusive assemblages of organisms. Selection, however, must in some cases act at the population level of organization. The intuitive argument of group selection proceeds as follows. Most species occur as demes, or subpopulations, the boundaries of which are set by the environmental mosaic. Often these populations are sufficiently isolated from each other to preclude much interpopulation migration and gene flow. Thus they operate as separate entities, each with characteristics falling within the continuum of variation possessed by the species as a whole.

Even if environmental characteristics are the same for all of these populations, some may become extinct while others prosper, since they themselves differ in response to environment. The more fit of these subpopulations should eventually recolonize patches of habitat vacated by their less successful counterparts. Therefore through differential probabilities of survival and extinction of entire populations, characteristics of the species might evolve. An example of this process concerns population age and reproductive structure (see Chapter 3). Some populations within a species may have few ages of reproduction, and they react by violent fluctuation in number to varying environmental conditions. Others that have more ages of reproduction are less reactive and more stable. Research using digital computers on which population dynamics can be simulated demonstrates that the reactive populations grow more slowly and have greater chances of eventual extinction than the demographically more conservative ones. Similarly, predators that overexploit populations of prey can eat themselves into eventual starvation. More prudent populations of these predators might eventually replace their greedier conspecifics.

There are two major objections to group selection theory. First, some supposed examples of group selection can be explained perfectly well in more conventional terms. Second, selection at the population level often seems to oppose selection of individuals. In the first case outlined above, the means to increased population stability and chance of persistence implies reduction of the reproductive potential and fitness of individual members of the successful population. Thus there seems to be a trade-off between these two modes of selection. Both may operate in nature.

GENETIC STRATEGY
Cost of selection

As yet intensities of selection have not been specified, but there appears to be a real problem with regard to this parameter. Estimates indicate that an average selection intensity of 0.01 per polymorphic locus might not be unreasonable. That is, for every locus that is polymorphic in a natural population, the probability of survival is 0.99. Recent work indicates that between 35 and 50% of loci may be polymorphic and segregating.

According to one method of calculation, which assumes that all loci are separate in terms of selection effects, the fitness of the natural population may be as low as 0.58×10^{-5}. Even more realistic models of selective death, for which selection is assumed to act on individuals rather than on genotypes, suggest that the adaptive strategies of populations must include ways either to minimize the selective death toll (genetic load) or to compensate for low viability fitness.

Darwin first realized the evolutionary importance of Malthus' conclusion that organisms often produce far more young than are necessary for population maintenance. He concluded that the excess births are the grist for evolution by natural selection, and he proposed the idea that only the fittest members of a population survive.

This simple observation may be quantified to some analytic advantage. Assume that $N_{t,0}$ animals are born to a population at time t. This number multiplied by the probability of survival (the viability fitness) gives the number of individuals in the population at the time of reproduction:

$$N_{t,r} = N_{t,0}(1 - d)$$
$$= N_{t,0}(\overline{w})$$

These survivors will each produce b offspring, so that if generations are discrete and not overlapping:

$$N_{t+1,0} = N_{t,0}(\overline{w})b$$

$\overline{w}$ stands for the total viability of the population taking selection at all loci into account. In order to persist any population must, on the average, reproduce itself. That is, $\overline{N}_{t+1,0} = \overline{N}_t$. Rearrangement of the previous equation shows that:

$$\frac{N_{t,0}}{N_{t+1,0}} = b(\overline{w})$$

The expected number of births per generation must balance the death rate. Regardless of what else is involved in the evolution of a population, the above condition must be met, on the average. As we will see later, if the product $b\overline{w}$ is greater than 1, the population could grow without limit, whereas a value less than 1 means eventual extinction for the population.

Apparently, about 99.5% of all species that ever existed failed to meet this primary requirement. Of those that are extant, most probably follow one of two fundamental strategies in recombination potential and heterozygosity. The strategies are not mutually exclusive; some species may follow intermediate roads to adaptive success. Strategies vary as the graininess of the environment varies.

Recombination. Species with high birth potential living in coarse-grained environments might be expected to have concave fitness sets. They will likely be highly polymorphic and will absorb high selection intensities, thereby remaining adapted to their fluctuating environment and reserving the ability to adapt to entirely new habitats. These species track changes in their environment by rapid genetic change.

Some insects are prime examples of such "trackers." High rates of genome recombination are an important aspect of such adaptive strategy; new gene combinations are continually necessary so that the populations meet the changing exigencies of their environment.

Several years ago Mather and Harrison studied the effect of recombination on the speed of selection. They selected laboratory strains of *Drosophila melanogaster* for number of abdominal bristles. Response to selection (increase and decrease in bristle number) was fast at first and then reached a plateau at which time change in bristle number was very slow. Selection was relaxed and lines combined for one generation and then selection was resumed. Resumption of selection brought about rapid response.

The results were interpreted to mean that recombination during the period of relaxed selection had provided new genetic variation upon which selection could act. In natural populations where environments continually fluctuate such periodic relaxation of selection combined with recombination-generated genetic variability must be important to the evolutionary process.

Heterozygosity. At the opposite end of the spectrum of adaptive strategy are the "weed" species. Much work shows that in these species only 15 to 20% of phenotypic variance is genetic; developmental homeostasis is of prime importance. Weeds are usually highly heterozygous. Remember that for biochemical reasons, heterozygotes may have wider tolerance to environmental variation than more homozygous individuals. It has become clear through the work of Stebbins and others that weeds maintain their heterozygosity by means other than balancing selection at the single gene locus. In fact, these species have drastically reduced frequencies of recombination. Many are apomictic (self-mating or vegetatively reproducing). This trait reduces recombination considerably, thus guaranteeing maintenance of existing (advantageous) phenotypes in the population. Most apomicts are faculative ones—cross-fertilization is a rare event, but when it occurs new genetic variability enters the population through mutation and recombination.

Oenothera has evolved an almost infallible mechanism for maintaining necessary genetic heterozygosity. *Oenothera* has 14 chromosomes and in many races all harbor translocations. Moreover, this set of translocations is so arranged that heterozygous members of the species are fertile and independent assortment of the chromosomes is impossible. Only one type of gamete is produced and this assures that each race of the species will breed true for its heterozygous condition. Thus high levels of heterozygosity, and presumably high incidence of coadapted gene complexes, are locked into races of the species.

Inversions. Inversions are of general adaptive significance and affect cost of selection. Because recovery of crossover products from inversion heterozygotes is very low, coadapted gene complexes, or supergenes, tend to develop. Loci affecting the same or functionally related traits tend to aggregate within linked groups. These ought to be arranged in such a way that complementary alleles occur on the same chromosome, guaranteeing maximal population fitness. Recombination of epistatically interacting alleles tends to lower fitness, since recombinant types are not optimal combinations. For example, shell color and banding pattern are closely linked in *Cepaea nemoralis,* a land snail that is selectively preyed upon by birds. Optimal associations of shell color and banding pattern give their possessor a high degree of protective coloration. Other combinations of color and banding are more visible to predators and are selected against. Genetic linkage lowers the incidence of unfit phenotypes.

Dobzhansky has studied the adaptive and evolutionary significance of inversion polymorphism in *Drosophila pseudoobscura.* This species has four major inversion types on the third chromosome, which are distributed geographically in a way that strongly suggests powerful adaptive significance and maintenance of various local population frequencies by natural selection. These inversion types go through striking seasonal changes in frequency. These changes and the fitnesses of inversion types may be related to changes in some environmental parameter like temperature. Anderson suggests that the chromosome types have different net fecundity distributions and thus may fit the model of selection suggested by Charlesworth and Giesel. According to this model the inversion type might be changing in response to changes in the population's rate of increase and age distribution (see Chapter 3). Whatever the cause of selection for chromosome type in the fly, it is important to realize that inversions represent structurally, functionally, and adaptively linked sets of alleles. The importance of such linked groups cannot be overemphasized.

Recently Franklin and Lewontin showed by computer simulation that structural chromosome abnormalities are not necessary for the establishment of linkage groups. They found that when chance of recombination between selectively significant loci with heterozygote advantage was varied, population fitness rises with recombination as alleles with similar selective value become part of the same homologous chromosome. Once advantageous associations form, they tend to be maintained by selection because recombinants have low fitness. Franklin and Lewontin also showed that once two loci become optimally related on a chromosome, they serve as a point of saltation for other nearby loci. This process of accretion of selectively related alleles on the same homologue can continue until most of the chromosome is involved. The effect should be particularly strong for loci that influence the same trait or traits which combine to form an optimal phenotype. Obviously, systems of major genes and their modifiers fit all criteria necessary for the evolution of selection-maintained interlocus correlations. Thus the dominance relationships of alleles with major phenotypic effects may be quite stable, particularly if the major and modifier loci involved are located within a correlated block of loci.

Ploidy and gene duplication. Essentially, a species evolves an advantageous genetic structure and sticks with it. A high level of fecundity is important to weedy species, for example, but in a different way from that in some insect and other highly fecund species. Weedy plants produce many seeds, but only a few fall in habitats suitable for germination. Mortality is concentrated among the seeds; once germination occurs, the probability of life until seed set is relatively great. Many weedy species are polyploid derivatives of ancestral species with normal chromosome numbers. The importance of this in terms of adaptation and genetic architecture involves the fact that polyploids have more than one genetic locus per enzymic function. Each polypolid duplicate of a chromosome carries duplicate, originally identical, genes.

Duplicate enzymes. Weeds and insects are only two examples of groups in which the individual's genetic architecture is important in determining an adaptive strategy in response to environmental heterogeneity. Recently, the rather important observation was made that in many species there are duplicate enzymes for the same biochemical task present in single individuals. These duplicate gene-enzymes often differ in reaction kinetics. The enzyme fructose dehydrogenase, for example, is duplicated in rainbow trout. One of the genetic loci determining this enzyme is active at low temperatures, the other at higher temperatures. Thus the range of temperature tolerance of the individual animal is increased by the different environmental ranges of the duplicated loci; genetic load, which is guaranteed by heterozygotes at a single locus, is avoided.

Deep water fish often have multiple enzyme systems or loci where their shallow water relatives have only one enzyme locus per function. The deep water species come from stable environments where physiological homeostasis might have been of evolutionary value. Such species often have greatly reduced recombination potential suggesting that stable environments might select for adaptations further reducing environmental grain. In contrast, shallow water species and others that live in widely fluctuating environments sacrifice developmental and physiological homeostasis for genetic flexibility. They have fewer duplicated gene complexes and higher recombination potential than those from stable environments. We will see later that genetic plasticity is very often of adaptive importance to species in ways other than those indicated here. In particular, such flexibility may be important to the species' ability to interact successfully with competitors.

Interpopulation gene exchange. Interpopulation gene exchange appears to be an important part of the adaptive strategy of many marine invertebrates, especially those occuring in the zone of shoreline that is alternately covered with water and exposed to air as a result of tidal action. In this intertidal zone, resident species are often faced with violently fluctuating environmental conditions from which they are largely unprotected. Mortality due to physical variation is often catastrophic. However, many species found in high intertidal temperate areas possess highly migratory larval stages that are capable of some interpopulation travel. The value of such larval stages is twofold. First, migratory larvae assure that no site with potential for the species will remain barren for long following extinction of a local population; planktonic larval forms are capable of resettling vacant sites on an almost continuous basis. Second, with migratory larval stages interpopulation gene exchange is possible. Such exchange has the same significance as suggested by Prakash and co-workers in their studies of *Drosophila pseudoobscura* (see p. 24).

Vermeij has compared geographic ranges of several species of marine intertidal invertebrates in light of where they could be found within sites, e. g. under rocks, high intertidal, low intertidal, in crevices, etc. His findings are summarized in Table 2-11. There is a direct relationship betweeen the amount of environmental heterogeneity experienced during the lifetime of individuals at one site and the degree to which the species is endemic versus cosmopolitan in distribution. Cryptic species, which are found in crevices in the rock, may be found over large geographic ranges as long as suitable crevices are to be found. These species are the ''weeds'' of the marine environment, occurring widely but associated only with special habitats. Deep intertidal forms have extensive geographic ranges for the same reason as cryptic species. The marine subtidal environment is relatively constant from time to time and place to place, at least in the tropics. High intertidal species, by contrast, are often endemic and occur only over relatively narrow geographic locations. Vermeij suggests that this occurs because high intertidal environments are subject to more geographic than temporal variation. Thus high intertidal forms must specialize on a microgeographic basis. Such speciali-

TABLE 2-11. Population structure and its relation to patterns of environmental heterogeneity

HABITAT	GEOGRAPHIC BREADTH OF SPECIES
Low intertidal, exposed	Moderately widespread
Low intertidal, cryptic	Widespread
High intertidal, cryptic	Moderately endemic
High intertidal, exposed	Highly endemic
Broadly distributed over intertidal zone, exposed or cryptic, most often exposed	Widespread

zation would require that there be little interpopulation migration. In fact, according to Vermeij, most tropic, high intertidal species have either no migratory larval stage or a much abbreviated one.

Vermeij's results do not seem to hold in temperate zones where seasonal fluctuations within an area may be as great as the variation among quite widespread geographic areas. Some high intertidal species are as widespread as their low intertidal relatives. Such species often have tremendously high levels of genetic polymorphism. One high intertidal barnacle is polymorphic at 85% of the loci that can be tested electrophoretically. It may well be that interpopulation migration serves to ensure that local genetic variation and adaptability are maintained.

Maximization principle. A fundamental theorem of natural selection and one of its extensions provide a basis upon which to begin the study of the ecological aspects of population biology. It is known universally as Fisher's fundamental theorem, and is named for the biomathematician who derived it. This theorem states that *the rate of increase in fitness with regard to a genetic trait is equal to the genetic variance of that trait*. Ecological traits for which there is genetic variability within a population can be selected. The extension, which is really of more importance to the population biologist than the fundamental theorem, states that *if two or more loci are being selected, then selection should change their allele frequencies in such a way as to maximize the rate of increase of population fitness, and presumably in such a way as to maximize fitness per se*.

This maximization principle is couched in terms of single genes, but there is no reason not to modify its context to include quantitative traits like height, or choice of food particle size, or wing length, or any of a vast number of other ecological traits. The importance of the principle is this: Any species or population will adjust to some change in its environment not simply with respect to that change, but also with respect to the change *in light of all the other environmental variables and selection pressures to which it is exposed*.

A physicist would think of the ecological species as a point moving in a vector field where forces act on it from all directions and with a variety of intensities. The direction and distance moved by this point in a potential field are determined as an average of all the forces at work. One representation of the population's potential field was described by Wright. Wright's concept was of an adaptive surface on which a population might travel to points of local maxima of fitness. We can think most easily of an adaptive surface in three dimensions. One of these we will designate as population fitness; the other two dimensions are genetic traits of ecological importance to the population (for example, high temperature tolerance and sun versus shade preference; alternatively, both of these traits might be related to the feeding habits of an animal). We assume that both of the traits have the requisite genetic variance associated with them. The expected long-term results of selection are shown in Fig. 2-14, *B*. According to this selection scheme, temperature tolerance has a larger associated fitness differential than does light preference. As a result, more change is expected in the range of temperature tolerance than in light preference. In terms of adaptive surfaces, the population is expected to climb to some peak of fitness relative to the two axes. (In actuality, the direction and extent of a population's adaptation to any one or a few environmental pressures are determined as well by selective pressures exerted by other aspects of the environment.)

Now look at the extended adaptive surface shown in Fig. 2-14, *A* with reference to population *X*. This population (and any others you might want to imagine on the adaptive plain) is assumed to be genetically diverse and thus to have a number of alternative avenues to higher fitness. Generally, a population will follow the path

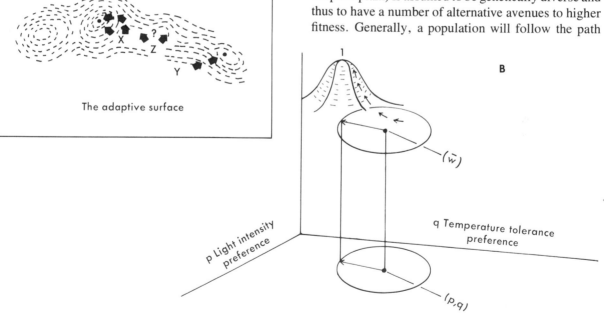

The adaptive surface

Fig. 2-14. The adaptive surface and maximization principle of natural selection. The population (in **B**) should change faster with respect to the characteristic that as a result of increased frequency will confer increased fitness more rapidly.

that offers the most immediate gains in fitness. Population X climbs the nearest adaptive peak, which also happens to be the highest one in the immediate area, whereas population Y climbs a near peak, but one conferring lower fitness than that reached by X. Should the environment change, thus changing the form of the adaptive surface so that some of the present peaks become valleys, these populations and population Z (which has a choice of adaptive strategies) will change again with respect to a multitude of genetic loci, thus adapting to current conditions. If changes in environment are sufficiently frequent, the environment becomes "fine grained" in Levins' terms. Gene complexes conferring broad tolerance ranges relative to important environmental parameters might then evolve.

Strategies and sampling

Founder effect. The founder effect enhances intrapopulation variability when combined with interpopulation migration. Each founder population should be a unique sample of the parental population. Since these populations are also unique recombinations of the parental population's genetic composition, they may harbor new phenotypes which might be of no value or even detrimental in the environment occupied by the original population. However, by colonization of new areas to which they may be fortuitously suited and by further recombination and selection, these variant populations should increase these species' phenotypic range of environmental competence. Thus colonization should be selected for in spatially heterogeneous environments.

Avoidance of inbreeding. Sampling effect can result in loss of intrapopulation genetic variability and adaptability. Therefore, it may be deleterious. In species that form small breeding populations, behavioral mechanisms to avoid loss of rare genotypes may be of advantage. Interpopulation migration is of value. There is evidence that some migratory birds which breed as limited subpopulations maintain variability because young do not return with any regularity to the area in which they were born, rather, they often join different populations.

In some species rare phenotypes have a distinct mating advantage. Ehrman demonstrated this in laboratory populations of fruit flies, and it may be general, although not so spectacular in natural populations. Blue geese have a mating advantage when rare in mixed flocks with snow geese. This again is an extreme example of a phenomenon which may involve more subtle phenotypic variability.

Genetic bottleneck. Some populations that occupy seasonally variable environments periodically are reduced to extremely small size (a sampling effect). If the loss of phenotypic and genetic variability and the changes in allele frequency which result are coincident with selection, increased rate of evolution can occur. Populations could conceivably adapt to changes in temperature more rapidly if temperature change is the cause of population collapse. However, loss of variability of other traits may be completely uncorrelated with the cause of collapse, and thus bottlenecks could also be disadvantageous. Theoretical considerations outlined here are almost completely lacking in development. However, if bottlenecks are deleterious to general fitness, balance might be provided by avoidance of inbreeding.

Age distribution and inbreeding: a special case. If populations are interoparous (contain more than one reproductive age class) their age distribution will change with changes in rate of population increase. As discussed in Chapter 3, declining populations will consist primarily of old individuals. Growing populations have large proportions of young. The effect of changes in age distribution on rate of loss of genetic variability has not been defined, but in numerically stable populations, differences in age distribution are responsible for large differences in rate of inbreeding. Populations with early reproduction and few reproductive age classes lose variability almost twice as fast as those with more spread out reproduction. Thus avoidance of inbreeding may also involve some optimal distribution of reproduction over age classes.

• • •

This chapter has presented the population genetic background you will need in order to continue your study of population biology. In the next chapter we will begin consideration of population ecology. For a time we will treat the population as a genetically monomorphic entity, but you should remember as you read that genetic diversity is the rule. You will be shown later that this diversity and natural selection are all-pervading aspects which determine and are determined by the existence strategies of natural populations.

BIBLIOGRAPHY

Baker, H., and G. L. Stebbins. 1965. The genetics of colonizing species. Academic Press, Inc., New York.

Bajema, C. J. 1971. Natural selection in human populations. John Wiley & Sons, Inc., New York.

Charlesworth, B. K., and J. T. Giesel. 1972. Selection in populations with overlapping gererations. II. The relations between gene frequency and demographic variables. Amer. Naturalist 106:388-396.

Crow, J. F., and M. Kimura. 1970. An introduction to population genetics theory. Harper and Row, Publishers, New York.

Dobshansky, T. 1958. Genetics of natural populations. XXVII. The genetic changes in populations of Drosophila pseudoobscura in the American Southwest. Evolution 12:385-401.

Ehrman, L. 1970. The matins advantage of rare males in Drosophila. Proc. Nat. Acad. Sci. 65:345-348.

Fisher, R. A. 1958. The genetical theory of natural selection, ed. 2. Dover, New York.

Franklin, I., and R. C. Lewontin. 1970. Is the gene the unity of selection? Genetics 65:707-734.

Giesel, J. T. 1971. The maintenance and control of a shell pattern polymorphism in Acmaea digitalis, a limpet. Evolution 24:98-120.

Giesel, J. T. 1971. The relations between population structure and rate of inbreeding. Evolution 25:491-496.

Giesel, J. T. 1972. Sex ratio, rate of evolution and environmental heterogeneity. Amer. Naturalist 106:381-387.

Kimura, M. 1968. Evolution at the molecular level. Nature 217:620-626.

Koehn, R. K., F. J. Turano, and J. B. Mitton. 1973. Population genetics of marine pelecypods. II. Genetic differences in microhabitats of Modiolus demissus. Evolution 27:100-106.

Kojima, K., and K. M. Yarbrough. 1967. Frequency dependent selection at the esterase locus in Drosophila melanogaster. Proc. Nat. Acad. Sci. 57:595.

Levins, R. 1968. Evolution in changing environments. Princeton University Press, Princeton, N. J.

Mather, K., and B. J. Harrison. 1949. The manifold effect of selection. Heredity 3:1-52.

Merritt, R. B. 1972. Geographic distribution and enzymatic properties of LDH allozymes in the fathead minnow Pimephales promelus. Amer. Naturalist 106:173-185.

Morton, N. E., J. Crow, and H. J. Muller. 1956. An estimate of the mutational damage in man from data on consanguineous marriages. PNAS 42:855-863.

Prakash, S, R. C. Lewontin, and J. L. Hubby. 1969. A molecular approach to the study of genic heterozygosity in natural populations. IV. Patterns of genic variation in central, marginal, and isolated populations of Drosophila pseudoobscura. Genetics 61:841-858.

Sheppard, P. M. 1951. Fluctuations in the selective value of certain phenotypes in the polymorphic land snail, Cepaea nemoralis. Heredity 5:124-134.

Slobodkin, L. B. 1968. Toward a predictive theory of evolution. In R. C. Lewontin, editor. Population Biology and Evolution, Syracuse Univ. Press, Syracuse, N. Y.

Stebbins, G. L. 1966. Processes of organic evolution. Prentice-Hall, Inc., Englewood Cliffs, N. J.

Thoday, J. M. 1958. Effects of disruptive selection: experimental production of a polymorphic population. Nature 181:1124-1125.

Vermeij, G. J. 1972. Endemism and environment: Some shore molluscs of the tropical Atlantic. Amer. Naturalist 106:89-102.

Wright, S. 1932. The roles of mutation, inbreeding, and selection in evolution. Proc. 6th Int. Cong. Genet. 1:356-360.

Wright, S. 1967. Surfaces of selective value. PNAS 58:165-172.

3 POPULATION DYNAMICS

Almost no natural populations maintain constant size. The literature of population biology is replete with examples of populations whose numbers fluctuate. Periodic fluctuations are reported for a variety of arctic animals such as voles, lemmings, arctic hares, and lynx. In temperate climates many insect populations are small in spring, grow rapidly through the summer, and decline again in fall. In the tropics similar species fluctuate consonant with seasonal changes in rainfall. Competition with and predation by other species often cause numerical changes in natural populations and may act to control their size (see Chapter 4).

Changes in population size are often correlated with changes in physical parameters of the environment such as temperature, water, light quality or quantity, and wind intensity. In addition, the numbers of individuals of a population are often related to changes in abundance of food or of nesting sites. All of these factors are components of natural selection.

Characteristics of population growth are central to the study of population biology. In terms of dynamics there are two or perhaps three different primary population structures. Simple organisms like bacteria, fungi, and protozoa have discrete generations: Individuals are born, reproduce, and die; there is no chance that a member of one generation will mate with an individual belonging to any previous or succeeding generation. More complex organisms, such as mammals, birds, fish, insects, and plants, usually have overlapping generations; a given female may re-produce more than once, and there is a chance that members of one generation will mate with those of subsequent or previous generations. Theoretical models describing this situation are called "discrete age class, overlapping generation" models. Models of continuously reproducing populations, with continuously overlapping generations, are algebraic extensions of this model. We will consider only the first two of these models, in turn. The overlapping generation model is a meaningful and important extension of the more simple alternative.

DISCRETE GENERATIONS

Many simple organisms exist in discrete generations. During each discrete time interval (Δt) a member of the population can die with a probability of $0 < d < 1$ and give birth with a probability of $0 < b < 1$. Thus, in a population of initial size N, the reduction in population size due to death in the time interval Δt will be Nd, and the size at the end of the interval is $N' = N - Nd$. The increment in population size due to birth is bN where b is birth rate. Therefore: $N' = N - dN + bN$ or $N' - N = bN - dN$ and $N_t - N_{t-1} = N_{t-1} (b - d)$. If birth and death rates are equal, $b - d = 0$ and the change in population size ($N_t - N_{t-1}$) is zero. The last of the above equations can be written in the form:

$$\frac{\Delta N}{\Delta t} = (b - d)N$$
$$= rN$$

where r is the sum of the birth and death rates, or the total rate of change of population size per time unit Δt, and ΔN is the change in population size, $(N' - N)$. This equation can be stated in words as: The change in population size that occurs over some interval of time, Δt, is equal to initial population size multiplied by the population's rate of increase. r can be considered as an interest rate. If b is greater than d, the population will grow; if the death rate exceeds birth rate, a decline in population size will occur. Table 3-1 shows a set of calculations giving N and ΔN for successive time periods; notice that although r is constant, ΔN increases with time. Actually, it increases as the population size, N, increases. This simply illustrates that populations grow at compound interest. Growth is dependent on capital at the time of growth, not on some

TABLE 3-1. Calculation of N and ΔN
for successive time periods*

TIME INTERVAL	INITIAL SIZE	CHANGE IN SIZE	FINAL SIZE
1	10.0	1.0	11.0
2	11.0	1.1	12.1
3	12.1	1.21	13.31
4	13.31	1.33	14.64

*The rate of growth of a population is equal to rN, where N is initial population size and r is the rate of increase ($= b - d$). Here r is 0.1.

TABLE 3-2. Growth of the human population

YEAR (AD)	POPULATION (IN BILLIONS)
1	0.25
1650	0.5
1825	1
1850	1.1
1927	2
1960	3
1975	4*
1985	5*

*Projection based on known growth rate.

more remote, past population size. Data collected for human world population illustrate this point strikingly (Table 3-2). The human population took until 1825 to grow from a few thousand to 1 billion. It took slightly more than 100 years for the addition of a second billion, 33 years for the third, and it will take only 15 years for the population to reach 4 billion individuals. Although it is undoubtedly true that $r = (b - d)$ has been increasing since prehistoric times (death rates have decreased drastically, especially in affluent societies), much of the accelerated growth stems from an increasing capital base.

This simple equation may be modified to a more general form by assuming that change in population size occurs over an infinitesimally small time interval, dt:

$$\frac{dN}{dt} = rN$$

which can be integrated to give $N_t = N_0 e^{rt}$. This final equation describes the size of a population at any time t after its inception as a function of its initial size, N_0, and its rate of increase r; e is the base of natural logarithms.

Fig. 3-1 shows the form of growth of a hypothetical population with birth rate $b = 0.6$ and death rate $d = 0.5$; r is $+ 0.1$ and initial size is 10. Notice that the population increases *geometrically*. Absolute change size is slow initially because N_0 is small, but as N_t becomes progressively larger, ΔN increases in magnitude. Under this model the population continues to grow without bound.

In the equation $N_t = N_0 e^{rt}$, r is termed the intrinsic rate of increase of the population, or the maximum rate at which the population will grow under any particular set of climatic or physiological conditions. Table 3-3 shows rates of increase of laboratory populations of the beetle *Calandra oryzae* at various temperatures and 14% grain moisture content. There appears to be an optimal temperature for which r_m is maximized. As temperature varies from this optimal value, rate of increase declines. The reasons for this variation in rate

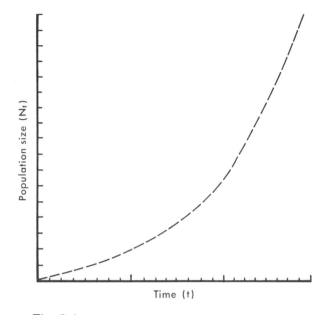

Fig. 3-1. Geometric population growth when $r = 0.1$.

TABLE 3-3. The rate of increase of *Calandra oryzae* related to rearing temperature

TEMPERATURE	INTRINSIC RATE OF INCREASE
23° C	0.43 per week
29° C	0.76 per week
33.5° C	0.12 per week

of increase are complex and will be treated in the following section when we consider such parameters as development time, fecundity, and survivorship distributions.

The maximal rate of increase differs for different populations within a species. Populations from cold northern regions generally have higher rates of increase at typical ambient temperatures than would warm weather populations if transported to colder regions. Temperature optima clearly depend to a large extent upon the conditions to which the population has

been exposed during its evolution. Since temperature has such an impressive effect on rates of population increase, we expect to find rather fine adaptation to this environmental parameter.

Extreme sensitivity to temperature should be most evident in insect and other poilkilothermic animals. Indeed, most of these forms breed only at certain times of year, which are governed largely by seasonal temperature differences. For example, *Drosophila melanogaster* has a single summer peak of population size, which occurs later in the summer in northern populations than in southern populations. Specialization has occured in *Ischnura ramburei*, a blue-green damsel fly. This species has only one reproductive period per year at the northern extreme of its range, where reproduction occurs only during summer. The insects spend the rest of the year as eggs and slowly growing nymphs. The situation is quite different in northern Florida, where weather during winter alternates between quite warm and moderately cool. In Florida predation of damsel fly nymphs occurs year round and water temperatures are always warm enough for larval development to continue. These Florida populations, where they have been studied, are multivoltine: many reproductive cycles are accomplished per year. At least one major reason for the population cycle differences between the northern and southern groups of this species relates to climate and temperature adaptation. Water temperature in Maine is probably too low for much growth of larvae in winter. Also, with heavy ice covers on damsel fly habitats, oxygen stores and food levels must be low. Therefore, growth and reproduction must be relegated to the summer months. In Florida, water temperature and oxygen content remain high enough throughout the year to permit larval growth and attainment of the adult form and reproductive maturity.

Another possible reason for this latitudinal difference in life history is more speculative. The Florida populations have many winter reproductive periods because death rates due to predation may be so high that periodic reproduction is necessary to replace winter populations. Why, then, isn't reproduction simply heavier in late fall or characterized by a single peak in midwinter? The amount of reproduction accomplished by the population in fall may already be maximal. Multiple rather than single midwinter breeding time may have evolved because of the selective action of temporal environmental heterogeneity. Periods of suitable weather in winter in northern Florida seldom last longer than 4 to 6 days. Damsel flies emerge as adults in response to warming trends. If they were to all emerge during a single warm period of insufficient length for successful reproduction, then extinction would be almost certain. When emergence is spread over many warm periods, at least one reproduction will probably be successful. This sort of discussion has been formalized into a mathematical theory of reproductive strategy known as the "relict seed phenomenon" (see p. 72).

In poikilotherms we might expect rate of increase to be affected by metabolic rate, and patterns of temperature optima to be the same for the two phenomena. This is true in many cases. Northern forms of a flounder-like fish, the Atlantic plaice, have very high growth rates when transplanted to warmer waters. The temperature-growth rate adaptation evidenced by members of the northern population results in low generation time, important to rate of increase.

Clearly, the above treatment of population dynamics involves many overly simple assumptions. A major case in point involves the earlier observation that populations do not grow at exponential rates without limit. Natural populations (perhaps other than seasonal insects) maintain remarkably stable sizes. Even seasonal insects, whose populations do grow exponentially at the beginning of the growing season, eventually reach numerical constancy or decline. Over the long term, their numbers remain relatively constant.

It appears that most natural populations are in some sense density regulated. For most populations the supply of some critical density-regulating resource eventually becomes short relative to demand as the population grows. Further increase in numbers is then limited by shortage of this requisite for population growth, or by the increased sensitivity of the more dense population to disease or predators.

Such density-regulating resources, or density-dependent factors, fall generally into three classes: (1) food, the availability of which can be further reduced by competition with other species; (2) space, usually that needed for nesting sites or foraging; and (3) predators, which increase their demands as prey increase in numbers. We will attack the problem of population regulation in far greater detail later, but for now we need only define the basic factors in order to proceed with our analysis of population demographics.

The above introduction to regulation can be summarized by a descriptive equation, which is really an extension of the exponential model of population growth. We begin with the familiar equation

$$\frac{dN}{dt} = rN$$

where r is the intrinsic rate of increase. To this equation is added a term to relate present population size to the number of animals of the species that the environment can support. The latter term can have many possible forms, but the relationship $(1 - N/k)$ is generally preferred. k is the carrying capacity of the environment, which is a difficult quantity to define. It is easy enough to state that any food source or area can support a certain number of organisms and that the maximum supportable number is the carrying capacity. This definition will do for now, but you will see later that other environmental factors influence to a significant extent the capacity of an area or resource to support life.

The final relationship, known as the Lotka-Volterra equation, is:

$$\frac{dN}{dt} = rN(1 - N/k).$$

As N increases, the fraction N/k approaches 1, and the term in brackets approaches zero. The population's realized rate of increase, r, can be written $r = r_m(1 - N/k)$. Thus as population size increases to k, r (the realized rate of increase) decreases to zero. These conclusions are shown graphically in Fig. 3-2; the population reaches a steady size characteristic of the environment's carrying capacity where r is zero. Should population size, N, exceed k, r will be negative and population size will decline. Change in carrying capacity can result in change in r and population size.

This is the simplest possible model of logistic population growth. It is a deterministic treatment where r must equal zero when $N = k$. More realistic modeling involves the probability statements of stochastic processes, ones that involve random fluctuations in relevant variables, number of individuals, and birth and death rates. With such models, r is most likely to be zero when $N = k$; but since it is allowed to be either positive or negative, k ceases to be an absolute maximum for population size. With such stochastic models for logistic or density-limited population growth, population number characteristically fluctuates about some central population number. Such changes are one possible explanation of cyclical changes in population size, which are characteristic of some species. Since natural populations can consist of phenotypes with

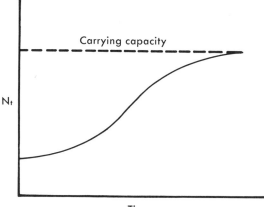

Fig. 3-2. Logistic model of population growth. Populations grow at their exponential rate until population size is half carrying capacity. Growth rate then declines.

different carrying capacities and rates of increase, maximum population size can also vary (within a population), depending on phenotype frequency. It is apparent that r and k are two factors peculiar to each species, and they are of evolutionary significance. For example, it would make no sense for a population that continually lives in an environment where space or food is limiting to increase its rate of increase. Ecologists, therefore, often talk about r selection, referring to selection for increased rate of increase, and k selection, referring to selection for more efficient use of some limiting resource. Both concepts of selection are perfectly valid and both probably apply to the same natural populations. Their relative importance is primarily a function of whether physical factor limitation or biotic limitation happens to be more intense.

OVERLAPPING GENERATIONS AND DISCRETE AGE CLASSES

When populations consist of individuals, each of which can produce young at more than one time during its life span, the basic variables of birth and death rate must take on extended meaning. Instead of the simple term d to stand for the probability that a female will die before reproductive age, we now define a set of values, d_x, to denote the probability of death before the start of each successive age or interval, x. Once we know the d_x distribution characteristic of a population under a particular set of conditions, we can calculate the population's survivorship distribution, l_x. This distribution is a set of probabilities that individuals born at time 0, at some one season or from some one birth cohort, will be alive at the start of each reproductive interval x. Of course l_x distributions differ markedly among species and individual populations. Data on this variable are presented for several species in Fig. 3-3.

Many species of animals or plants could have been chosen to contribute to this illustration. The ones shown represent well-analyzed cases or various kinds of survivorship distribution particularly well. Survivorship distribution A is compiled from data collected on the Dall sheep population from Mount McKinley National Park in Alaska. The distribution begins with 100% at age zero. This simply signifies that all individuals belonging to a birth cohort are assumed to be alive at birth. In this particular population survivorship after the first half year is 0.60 and by the end of the first year only 28.5% remain alive of the original group born at the beginning of the year. The investigator determined that infant mortality was the result either of predation

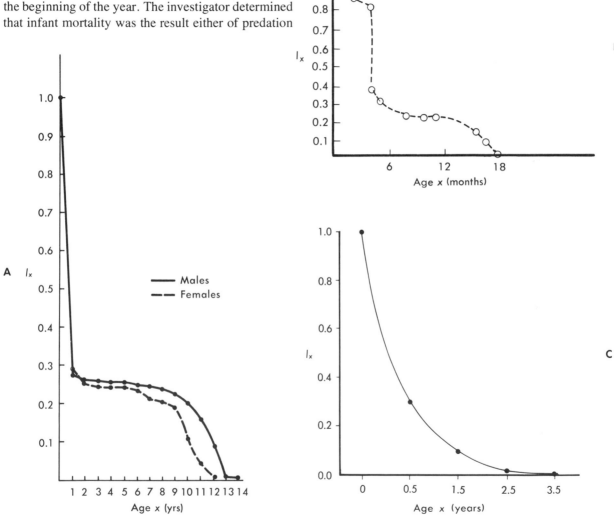

Fig. 3-3. Survivorship curves for **A,** a population of Dall sheep on Mount McKinley; **B,** the barnacle *Balanus balanoides;* and **C,** the ring necked pheasant. Since the sheep population **(A)** was stable, first-year survival is adjusted to give r ≈ 0.

or of accidental death, probably largely during migration and the first winter.

The distribution flattens out following the first year of life; only about 1% of the initial cohort dies during each year up to age 11. At about age 11, the death rate increases drastically. This is thought to be because of old age and senility, which lead to debilitating diseases. Such a survivorship distribution appears to be nearly ideal considering the conditions under which Dall sheep live. Early death accomplishes two things. First, most selective mortality occurs *before* the age at which sheep are reproductively mature. Thus natural selection for agility, resistance to disease, extremes of weather, intelligence, predator avoidance, and feeding efficiency can be maximally effective. Very few marginally fit animals survive to procreate. Second, since those sheep that die do so at an early age, less critically short resource is expended on animals that are destined to contribute little to the fitness of the population. Finally, every female that attains reproductive age can expect to live through most of the ages of reproduction. Net reproduction is nearly equal to gross or physiologically realizable reproduction. The importance of this latter point will become clear soon. Here we have an example of r and k selection complementing each other; death of young might be due to accident, but survivors are less food limited than if no young had died.

Distribution B is quite different from A. The barnacle *Balanus balanoides,* a sessile (permanently attached) animal, is quite common in the rocky intertidal zone along the ocean's edge. Mortality during the first 2, planktonic weeks of life is immense. It is thought that only about 0.005% of all the fertilized eggs survive larval life to settle on the intertidal rock faces. The survivorship distribution shown here is begun at the time of settling. Reasons for this high death toll are manifold; barnacle larvae are not particularly motile and hence are unable to avoid capture by predators. Many, undoubtedly, fail to find food, and still others drift out of the influence of near shore currents and are lost at sea.

During the first 6 weeks after a group of young barnacles settle from the plankton, an additional 70% die. The reason for this extensive death is somewhat obscure, but because of their long planktonic larval life barnacles are capable of rather extensive travels. A barnacle conceived along the North Carolina coast may travel to Florida or perhaps to Martha's Vineyard before settling. This, of course, means that many young barnacles may have settled in areas remote from those to which they are best adapted. Much selective death could be the result of such intense migration and population admixture. In fact, in my laboratory I am finding that rather remarkable changes in population gene frequency occur during the first week or so after a group of barnacles has settled from the plankton. Barnacles live for a few months, and this is reflected by the continual low rate of survivorship illustrated in Fig. 3-3, *B*. Survivorship distributions such as those of the barnacle are suggestive of a rather harsh environment and are characteristic of species that exist in environments that are extremely heterogeneous and coarse grained. They are also associated with intense migratory activity, the strategic basis of which is genetic adaptation and adaptability to spatially and temporally heterogeneous environment. (Recall the discussion of *D. pseudoobscura* in Chapter 2.) Very young barnacles are generally poor at physiological acclimation.

Fig. 3-3, *C* is the survivorship distribution of the ring-necked pheasant. Although early mortality is somewhat higher than later death rates, death rate in general is rather constant from age to age. Such data exemplify the so-called "accidental" type of survivorship distribution. Constant proportions of animals die per age class. Such distributions are quite general among birds, and together with the curve illustrated by the Dall sheep, may be characteristic of species from fine-grained environments in which resource shortage, infectious diseases, and inclement weather are the major causes of death.

Historically, man's survivorship distribution has

TABLE 3-4. Percent survival of males and females in the United States from birth to ages 15, 30, and 45*

YEAR OF BIRTH	AGE 15		AGE 30		AGE 45	
	FEMALES	MALES	FEMALES	MALES	FEMALES	MALES
1840	66.4	62.8	58.1	56.2	49.4	48.2
1880	73.1	71.5	67.4	65.7	61.1	58.3
1920	89.8	87.6	88.0	83.4	85.8	79.8
1960	97.5	96.6	96.9	95.1	95.9	92.9

*Data from Kirk, 1963. Cited in Bajema, C. J. 1971. Natural selection in human populations, John Wiley & Sons, Inc., New York.

varied markedly. Table 3-4 shows selected data for the United States. Survivorships for the birth class of 1960 are projections and are already proving to be low estimates. Death rates related to infectious and epidemic diseases, many of which occur early in life or at least before reproductive years have passed, have been lowered through modern medicine and public health measures. There seems to be no evidence of acquired genetic resistance to various diseases like plague, pneumonia, diarrhea, influenza, and the significant killers of the nineteenth century. This is reflected by changes in death rates shown in the tables. Much of man's death is now due to organic, degenerative diseases such as cancer and heart disease, which generally do not kill until after completion of reproduction. For this reason, these diseases have little effect on the human population's rate of increase, and little genetic progress relative to these causes of death is likely.

When we consider in more detail the distribution of causes of human death, we cannot ignore one somewhat unsettling fact: 33% of the female deaths before age 30 and 60% of those for males are now caused by congenital malformations and diseases of early infancy, many of which involve genetic defects. One wonders what effect increased dosage of ionizing radiation and perhaps chemical pesticides, treatments, additives, etc., are having, and whether these various pollutants are not at least partially responsible for such genetic death. Of course, poor diet and lack of prenatal care undoubtedly contribute to early death and may

also enhance the mortal expression of suboptimal genetic structure.

The data of Table 3-4 are somewhat unfair representations of death in the United States since if survival were computed from conception rather than from birth, there would be a significantly lower survival rate. Much fetal wastage and stillbirths, often of malformed fetuses, is due to genetic abnormality. Those genetic abnormalities that are medically intractable are constantly being removed from the human population. But even if mortality of congenital defectives before the age of first reproduction is 100%, those abnormalities that are recessive in character have a long life time in populations such as ours. In fact, from a theoretical point of view, complete removal of phenotypically recessive mutants is almost impossible. Death rates are much higher for underdeveloped countries. One factor causing the survivorship differential between the United States and less advanced nations is related to carrying capacity—specifically, the quality and quantity of food available. This is particularly important among the lower social classes in underdeveloped countries. Much infant death is due to crowded conditions which enhance the effects of both nutritional deficiency and infectious disease.

The distribution of survivorship over age classes becomes important when viewed in light of fecundity distributions. Combinations of two life table components, which confer higher fitness (in the sense of rate of increase), survivorship, and fecundity, are continually selected for in natural populations.

THE m_x OR FECUNDITY DISTRIBUTION

How do survivorship and age-specific probability of birth interact to determine rates of increase? It is obvious that the survivorship distribution per se is of no importance to a species if reproduction all occurs at one age. However, most animals other than protozoa are iteroparous (breed more than once during their life span), and natural selection has led to a variety of timings of reproduction. We will consider several of these modifications in reproduction, as they are manifestations of natural selection producing strategic modification for living in a variety of environments.

The m_x distribution states how a species distributes its reproductive effort over the life span of its adult members. Fecundity distributions vary greatly from species to species; two examples are shown in Fig. 3-4, *A*. Fig. 3-4, *B*, shows the interaction of these distributions with a survivorship distribution assuming "accidental" or age-independent death. Fig. 3-4, *B*, is reproduced in Table 3-5 for the two extreme distributions (where most reproduction is early versus where each female has the same total number of offspring, but has equal fecundity for all age classes). Gross repro-

duction is the same for both hypothetical populations, but rates of increase are vastly different.

In order to calculate intrinsic rate of increase, two derived values must be considered: (1) net reproduction per female, or the number of female offspring the average female is likely to have during her lifetime considering that some of the population's females are

TABLE 3-5. $l_x m_x$ distributions of the two populations shown in Fig. 3-4

AGE (x)	$l_x m_x$ (1)	$l_x m_x$ (2)
1	0.5	0
2	0.25	0
3	0.25	0.25
4	0.25	0.125
5	0.125	0.3125
6	0.0325	0.0975
7	0	0.01626

$$\sum l_x m_x = R_0$$

$=$	1.4075	0.80125

$$T = \frac{l_x m_x x}{l_x m_x}$$

$T \quad =$	2.536	4.382
$r =$	+0.135	−0.051

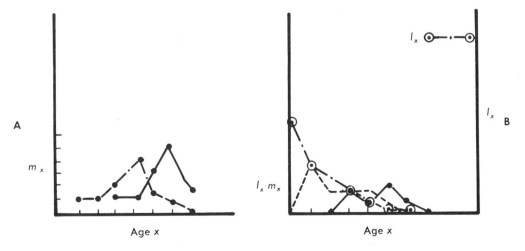

Fig. 3-4. **A,** Two hypothetical distributions of age-specific fecundity. **B,** Interactions of m_x with an accidental death type l_x distribution.

likely to die before producing one, or two or more times, and (2) the generation interval, or average age of reproduction, which simply expresses the time taken by the average female to produce her offspring. We can use the data in Table 3-5 to illustrate how to calculate generation interval, T, and net reproduction, R_0, and from these calculate r. Both calculations are numerically and conceptually simple. R_0, the net reproduction per female, is the number of offspring produced in a breeding cycle by all females alive during that cycle divided by the total size of the female population during the cycle. In mathematic terms this quantity is given by

$$\frac{\sum\limits_{x=1}^{d} l_x m_x N_{0_{t-x}}}{\sum\limits_{x=1}^{d} l_x N_{0_{t-x}}}$$

the summation of numbers of births per age class multiplied for each age by the number of females of that age, $N_{0_{t-x}}$, where $N_{0_{t-x}}$ is the number born $t - x$ age intervals ago. When the population is of constant size $N_{0_{t-x}}$ cancels out and:

$$R_0 = \sum\limits_{x=1}^{d} l_x m_x$$

where d is the age of last reproduction. T, the generation time, is the average age at which the population's females produce young. The quantity is calculated from:

$$T = \frac{\sum\limits_{x=1}^{d} l_x m_x x}{\sum\limits_{x=1}^{d} l_x m_x}$$

Thus T is the weighted average age of reproduction, where weighting of ages (x) is on the basis of numbers of young produced by females of each age x.

How do these quantities relate to the intrinsic rate of increase? Consider the equation $N_t = N_0 e^{rt}$ and make t equal to the interval of one generation, T; then we have

$$N_T = N_0 e^{rT}$$

From the definition above, N_t/N_0 must be equal to R,

so:

$$R_0 = e^{rt}$$

Taking logs and solving for r, we get:

$$r = \frac{ln R_0}{T}$$

Thus a population's intrinsic rate of increase is a rather complex function of its age-specific birth and death rates. Although r is still the interest rate at which populations grow, this quantity is more clearly and operationally defined as a function of the number of young produced by females and the average time elapsed between the birth of a female and the birth of her young. This relationship is shown numerically in Table 3-5. Both populations have the same gross reproductions per female, but the population with a short generation time has a rate of increase of 0.135; r for the population that reproduces later is -0.051.

At high rates of increase the above equation is grossly inaccurate. More accurately, r is the root of $1 = \sum\limits_{x}^{d} e^{-rx} l_x m_x$. It can be obtained by iterative calculation.

These two example populations and the above calculations lead to several conclusions: Rate of increase is determined by the relation between a female's probability of survival to any age x and the number of offspring she can produce at various ages. Early reproduction results in increased net reproduction (relative to gross reproduction) and in decreased generation time. Since r is measured in terms of increase per unit time rather than per generation, this latter effect of early reproduction will result in a higher rate of increase. Populations with the same amount, but late reproduction will obviously have lower rates of increase, because each typical female takes longer to produce her requisite number of young.

COMPONENTS OF DEMOGRAPHY AND WEATHER

Components of weather, such as temperature and humidity, should have profound effects on development time, death rate, and rate of fecundity of animals, particularly poikilotherms. Data collected on a locust population by A. G. Hamilton is presented in Fig. 3-5. There are distinct optima for temperature. The four parameters depicted in this figure are those which define gross fecundity and development time. There are apparently no data available on survival at the various temperatures and humidities. If there were we could at least approximate r. But inspection allows us to estimate the effects of temperature difference. For example the difference in development times between 28° and 32° C at a vapor pressure of 40 mm Hg is

roughly 8 days. In a species with a minimum egg to adult development time of 30 days, this difference is tremendously significant even when no information on survival is available.

DEMOGRAPHY AND ADAPTIVE STRATEGY

Parameters of life table distributions can be naturally or artificially selected. In flour beetles and fruit flies age of first reproduction (development time) can be changed by artificial selection. Development time and the form of the m_x distribution (early, late, or evenly distributed reproduction) vary from species to species and even in separate populations within species.

Although we will investigate several specific examples of adaptive strategy related to demographic structure, this is a good time to lay a general framework, which depends in large part on Levins' concept of the fitness set and adaptive strategy. Levins distinguished two rather distinct kinds of species, those from fine-grained environments and those from selectively significant, coarse-grained ones. Those species from coarse-grained environments, where grain is related to physically selective variables, ought to have high death rates or be subjected to an unpredictable set of environments. Furthermore, death might most often occur to young when genetic variability and proportion of marginally fit phenotypes must be highest of all classes. Such species may be characterized as having a barnacle-type survivorship distribution, and they ought to compensate with early, massive reproduction.

Since environmental grain is a function of how organisms "see" fluctuations, smaller forms, with their relatively poor acclimation ability, will be more affected by small environmental perturbations than larger species. In other words, small animals, like herbivorous insects, live in a coarser grained environment than do larger ones. Therefore, such species ought to have evolved high birth rates and early first age of reproduction in compensatory response.

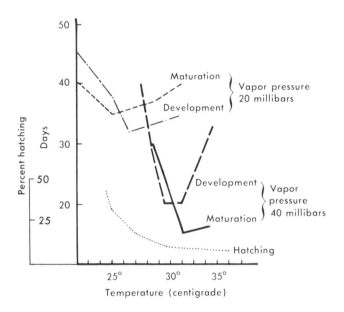

Fig. 3-5. Effect of air temperature and humidity on hatching, development, and maturation times of an insect. The two environmental variables interact on all of the rates. At high temperatures the interaction produces a single stress of water loss, which is being measured. Effects of the same variables are shown on egg hatching success. Hatching success is maximized (44%) at 25° C. These effects produce changes in population rate of increase. (Data from Hamilton.)

Relation between life table functions and age distributions

According to the previous discussion, the interaction of a population's life table functions defines the rate at which a population can grow in an unlimiting enviornment. This says nothing about the number of individuals that occupy various age classes of a population at any one time. The form of population growth and fluctuation in numbers depends on the instantaneous rate of increase, or the relationship between numerical age distribution and the fecundity distribution. The number of individuals in a birth class depends on the number of adults alive at the various reproductive ages born $t - x$ ages ago, multiplied by the probability of these individuals still being alive at the present time and the number of births characteristic of each reproductive age class. $N_{0,t-x} l_x$ defines the number of individuals in the xth age class. This, multiplied by the probable number of births per female of that age gives the number of offspring contributed by females of age x, and summation over all ages gives the size of the next birth class.

As a population grows or declines in size, the number of individuals in its various age classes fluctuates. As a general rule, if a population has an intrinsic rate of increase of zero, $N_{t,0} = N_{t-1,0}$ the numbers of individuals in its age classes remain constant, and their proportions are the same as the l_x function. A shift to a positive rate of increase results in proportional accumulation of young individuals, because more young are produced at each generation. Similarly, once the population has reached stable age distribution, shift to a negative rate of increase results in a population composed largely of old adults, because fewer young are born at each successive breeding interval.

Fig. 3-6, a, represents a growing population. Most individuals are young, suggesting that there has been a recent high rate of increase of reproductive individuals. In terms of future growth, the large class of young individuals will eventually enter reproductive life and, if they are as fecund as their parents, the population will continue to grow rapidly. Fig. 3-6, b, represents a stable population. Age classes are present in nearly equal frequencies. This pattern suggests neither growth nor decline. Fig. 3-6, c, is characteristic of a declining population. Since few individuals are being born, a greater proportion of the population is old. Such a population should continue to decline because there is no numerous class of young to contribute to a future period of high population growth.

A final hypothetical age distribution is shown in Fig. 3-7. Such an age distribution is interesting be-

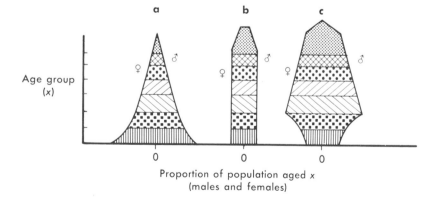

Fig. 3-6. Age structures characteristic of a, rapid population growth; b, a stable population; and c, a declining population.

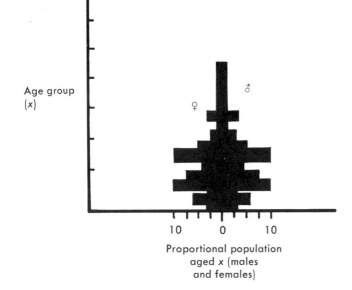

Age group
(x)

♀   ♂

10   0   10

Proportional population
aged x (males
and females)

Fig. 3-7. A hypothetical age distribution implying violent past shifts in birth or death rates.

cause speculation can be made about its cause and consequences. Proportions of individuals shift violently from age class to age class, and there is no clear progression from young to old. Clearly this distribution is the result of recent drastic fluctuations in birth or death rate. The distribution's broad points might be interpreted as reflecting periods of high birth rate. Less well-represented age classes could denote periods of low birth rate or, possibly, years of high age-specific death rate (for instance a polio epidemic). The best prediction that can be made for this population is that until the age classes pass through reproductive age, age structure and population increase rate will continue to fluctuate. In fact, the effects of this irregular age distribution might be felt for a number of generations, because the illustrated waves of population size and age distribution will tend to be propagated. They should be damped only slowly.

Age distribution effects are one reason demographers predict that even if we were to achieve a birth rate consonant with zero population growth in the near future, actual zero growth will not be achieved, in a stable sense, for at least 70 years. The present age distribution of the United States is quite irregular with some large and some small pre-reproductive age classes. The actual growth rate of our population will continue to fluctuate. Zero growth will be obtained only when the age distribution reaches a stable form similar to that of example b in Fig. 3-6.

The human age distribution (Fig. 3-8) of an area of the West Virginia Appalachians is interesting. Large segments of this population are either beyond child-bearing age or are below the age of 15. Such an age distribution might indicate a drastic drop in population birth rate at some time in the population's recent history. Without the large class of pre-reproductive individuals we would certainly have to conclude that this population is declining. In fact, the 0 to 5-year-old age class is quite a bit smaller than the 5- to 9- and 10- to 14-year-old classes. However, this age distribution actually reflects heavy migration from the population. This migration is of individuals of child-bearing age and is causing a drop in population size and rate of increase. It also results, in this case, in an economically unbalanced population. Large proportions of the population are either too young or too old to work.

By now, you should realize that there are two ways in which our population's rate of growth and eventual size may be controlled. The most obvious of these has been widely publicized. Considering our population's current death rate, it has been calculated that if live births were limited to 2.1 per couple, the eventual rate of increase reached would be zero. This is a long-term process.

A second strategy stems from the observation that age of parents, both in terms of R_0 and of T (the generation interval), strongly influences r. The data in Fig. 3-9 illustrate the underlying principle graphically. In calculating the numbers presented here, we assume that r should be 0.02. Compared are the average ages

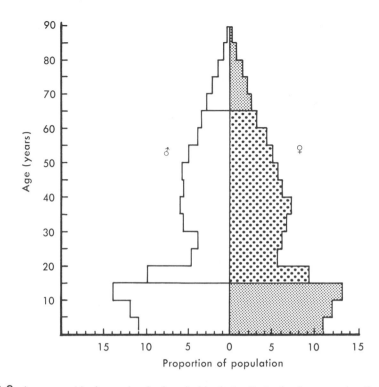

Fig. 3-8. Age pyramid of a region in Appalachia that reflects the heavy emigration of the younger elements of the reproductive age group. This emigration results in a high ratio of dependent young and old to the economically active. It also reflects a declining population in the region. (From Smith, R. L. 1972. The ecology of man: an ecosystem approach. Harper and Row, Publishers, New York.)

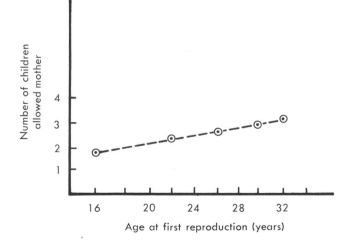

Fig. 3-9. Number of children allowed females who begin reproductive life at various ages in order to give a rate of population increase of 0.02 per year.

of reproduction and net increases that will produce a rate of 0.02 per year. It is assumed that a mother produces her children at 2-year intervals. Notice that the average female who has her first child at the age of 16 years contributes almost twice as much to population growth as a woman who waits until the age of 32 years to have her firstborn. By postponing age of first reproduction it is possible to have relatively large families and still be part of a population with a low rate of increase. However there is a catch; the number of deleterious mutations and birth defects increases with the mother's age. A commonly cited example is Down's syndrome (mongolism).

Data on the reproductive habits and rates of increase of several additional species serve to emphasize the importance of number of young versus the age at which they are produced. In Table 3-6 are data on R_0, T, and r_m for two mammals and two insects. Although the Norway rat produces as many offspring as the mouse, the mouse's rate of increase is greater because its generation time is shorter. The same point is made even more dramatically by comparing the louse and the grain beetle.

One relationship between fecundity and survivorship distribution is easily seen. Those species for which early mortality rates are high must have large numbers of young early. For these, commonly referred to as colonizing species, development time should be short and number of young should be large. Species like barnacles reproduce after only 1 or 2 months of life and release large numbers of young into the water.

Several species of limpet (snails that live in the extremely harsh high intertidal of the marine environment and suffer high early mortality) spawn huge numbers of eggs. In general, high mortality often leads adaptively to high fecundity and short development times. However, it is often difficult to assign cause and effect to such variables.

Although high mortality may be compensated by intense reproductive effort, reproductive effort often leads to high mortality. A study done in 1957 on population dynamics of some California populations of black-tailed deer shows very well how fecundity and survivorship are related to environmental carrying capacity. This investigation was pursued in two areas, a shrubland, which was thought to have both abundant cover and food, and an area of chaparral where both food and cover were less abundant. In the shrubland an average of 1.65 fawns were produced per doe whereas fawn production was only 0.77 per doe on the chaparral area. These differences were attributed to more abundant food in the shrubland area. Survivorship distributions were gathered from census data for each of these areas (Fig. 3-10). Death rates of females were higher on the shrubland than in the chaparral. The investigators attributed this difference to the higher fecundity characteristic of the shrubland population. Why do you suppose this is true?

The answer to this question lies in the fact that producing offspring is a relatively dangerous undertaking for an adult animal. Does carrying fawns are heavier and less agile and, therefore, are more susceptible to predation and accident. In addition, fawns are carried through the winter months when quantities of available food are at their yearly minima. Does with in utero young must feed more actively than those with no young, and the energy requirements of those with twin fawns are even greater. Starvation becomes a greater problem for more fecund does than it is for those carrying either one or no young. Thus the higher death rate of shrubland does can be attributed to more intense physiological reproductive drain on them than on the chaparral does.

TABLE 3-6. Net reproductive rate (R_0), mean generation interval (T), and intrinsic rate of increase for four common animals

ANIMAL	R_0	T		r_m
Norway rat	25.66	217.57	days	0.0147
Mouse	5.904	141.75	days	0.0125
Flour beetle	275.0	55.6	days	0.101
Human body louse	30.93	30.92	days	0.111

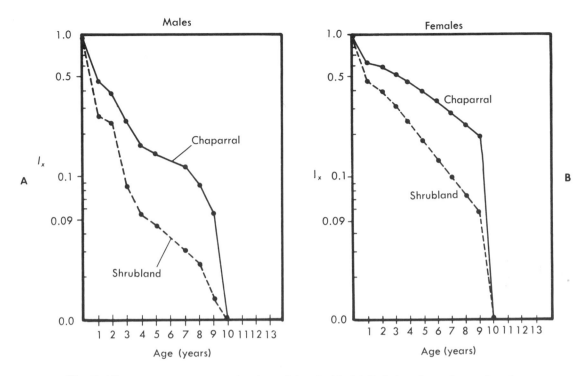

Fig. 3-10. Survivorship curves of male and female black-tailed deer from chaparral and shrubland. (Excerpted from Taber and Dasmann, 1967.)

There is a rather delicate balance between survivorship and reproduction; high rates of fecundity tend to reduce survivorship. As might be expected, components of this balance are subject to natural selection. In a recent study on dandelions, investigators found two subpopulations. One group occupied an area that was mowed once per year to a height of about 8 inches; the other area was an open lawn that was mowed frequently. The unmowed area, located along a creek bank, was inhabited by perennial plants and provided an intensely competitive situation for the dandelions. The lawn area was occupied by annual and perennial grasses and was subject to heavy pedestrian traffic. Thus the lawn was a site of *r* selection (heavy density-independent mortality), whereas the creek bank was characterized by little catastrophic mortality, but heavy interspecific competition (see Chapter 4).

Members of the open lawn population were devoting a far greater proportion of energy and plant structure to flowers than to vegetative growth. This was not true of the members of the seldom mowed population. The mowed group, because flowers were constantly destroyed, had at least partially evolved toward devoting more energy to reproduction, hence faster growth of flowers and more certain propagation. The creek bank population, by devoting more energy to leafy structure and tall seed-bearing stems, apparently increased its ability to crowd out the competitors and to disperse seeds from within a veritable forest of surrounding tall grass stems.

STRATEGIES OF NET FECUNDITY DISTRIBUTION

The net fecundity distribution of a population is obviously one of its more important adaptive attributes, and we expect that the variation observed between and within species might be related to the particular set of environmental problems which the species or populations have faced in the past. Thus a body of ecological and evolutionary theory has grown up to explain in general terms the kinds of variation we observe in net fecundity distributions.

As you read the following pages, you should consider that no species follows any particular strategy of net fecundity distribution exactly. Because a population's environment is constantly changing to favor various strategies in turn, observed net fecundity distributions may represent the result of trade-offs in adaptive value, or they may reflect intermediate steps in the evolution of some strategy. Thus the strategic conclusions we are about to discuss are broad generalities designed to explain broad categories of populations, not specific ones.

The three types of species we will look at with regard to reproduction are (1) the colonizing species, (2) species which are limited by resources or variable chance of reproductive success, and (3) relic species, which are apparently fighting a rear guard action in a deteriorating environment.

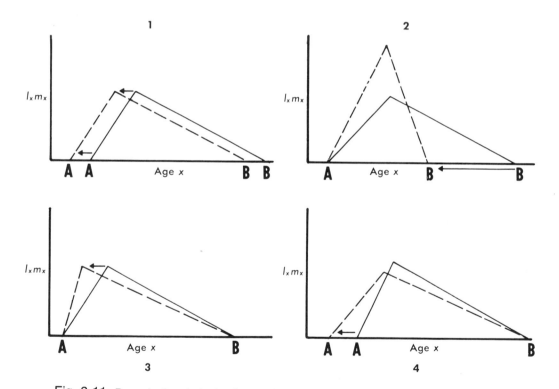

Fig. 3-11. Reproductive strategies that can increase a population's intrinsic rate of increase. **1,** Rigid shift of v_x (shift A, B, and T [peak age of reproduction]); **2,** shift B only (more compact v_x schedule); **3,** shift T; B and A remain constant; **4,** shift to earlier age of first reproduction (A only).

Colonizing species—species from temporally coarse-grained environments

Apparently, there are two basic types of colonizers, insects and weeds. Not all "weeds" are true weeds and not all "insects" are phylogenetically insects. The adaptive strategies of weeds are discussed in Chapter 2. "Insect" refers to a species that continually adapts to changes in environmental parameters (see Chapter 2). These species need high rates of increase. There are a number of basic strategic ways this can be accomplished. Earlier age of first reproduction (reduction of development time), earlier average age of reproduction (left shift of the m_x function), increased total number of offspring, and increased length of reproductive life are all ways by which r can be increased. Fig. 3-11 shows these strategies graphically. A is age of first reproduction, T is age of greatest reproduction, B is the last age of reproduction and R_0 is the area under the reproductive triangle, which is equal to $\sum l_x m_x$. R. C. Lewontin first showed how to analyze such distributions for reproductive strategy. His most pertinent results are shown in Fig. 3-11. r can be most expeditiously increased from some initial value by reducing all three values, A, B, and T. A reduction of 1.55 days has the same effect on r as a 2.2-day shift in age of first reproduction or a 21-day increase in length of reproductive life. All three changes in the species' fecundity distribution change both R_0 and generation interval, but reduction of generation time is of far greater importance to r than is an increase in R_0 of similar magnitude. If development time is shortened, T is maximally reduced, whereas increase in length of life *increases T*.

If the above information is considered in light of the maximization principle of natural selection, it should become obvious that the best way to increase r for colonizing species is reduction of development time. This strategy involves the smallest and presumably easiest change in a population's fecundity distribution; hence it should result in the fastest increase in

population fitness. This sort of manipulation of fecundity distribution seems to be general in species that are not food limited and that live in fluctuating environments.

Most seasonal insects produce large numbers of eggs early in life and have relatively short development times. Perhaps parthenogenesis (reproduction without fertilization) is merely an extension of this sort of strategy. Some insects, such as aphids, colonize their environments each spring; and some temporary pond organisms, such as species of rotifer and *Daphnia,* hatch from diapause eggs and produce only females, which then produce new females without mating in the spring. These species produce male and female offspring as conditions worsen in the fall or as temporary ponds begin to dry up or as population density increases. Among invertebrates, developmental plasticity and poikilothermy may interact to maximize r. In poikilotherms metabolic rate is a direct function of body temperature, and high ambient temperature and body temperature result in high metabolic and development rates, hence short development time. A warming environment often corresponds to the growing season for these organisms, so shifts to short development time are automatic and physiological for these species. Perhaps this observation helps to explain why poikilotherms and invertebrates in general have enjoyed great evolutionary success and are far more common than homeotherms.

Very little concrete data exist, but as we learn more about insects we will probably find that those with short development time or early reproduction are the most successful at adapting to catastrophic environmental changes. We already know this with respect to broad categories of insects. Predatory forms with low rates of increase are harder hit by general pesticides than are their more fecund prey, the pests.

If one compares extremes of environmental heterogeneity with earliness of reproduction in lizards, one finds that those species living in physically rigorous environments (desert and temperate zone) have significantly earlier reproduction than do those from physically more benign environments. Similarly, those lizards that are heavily preyed upon have early development, but predatory forms take longer to develop and have fewer offspring.

Conservative strategy

The second type of fecundity distribution for colonizing species is perhaps the most common among natural populations. It might best be named "conservative." Instead of reproduction being concentrated among a few earlier age classes, reproduction is spread out over a large number of age classes. Such fecundity distributions are well exemplified by various fishes, some of which are represented in Table 3-7. Notice that these fish have delayed reproduction and long reproductive lives; in addition length of reproductive life increases with environmental variability, of which variation in spawning success is an indicator. This environmental variability refers to variation of conditions that somehow determine the success of the reproductive season. Why is it that fish found in reproductively more variable environments have more attentuated reproductive lives? The answer to this question involves the number and severity of relatively unsuccessful reproductive years that a population must weather before a favorable one occurs. Assuming that "bad" years are years in which no reproduction occurs, then any population that survives such an environment must have had at least one more reproductive age than the longest string of bad reproductive years.

Of course, it is unrealistic to assign the categories "good" and "bad" to reproductive years. It is, however, quite reasonable to ask what sorts of population fecundity distributions are most favorable in fluctuating environments. In a series of computer simulations of theoretical populations with various types of fecundity distributions, populations were exposed to artificial environments that produced changes in R_0 (hence in realized rate of increase, since the populations were never near stable age distributions at which intrinsic rate of increase applies). The average realized rate of increase of populations with large numbers of reproductive age classes increases as environmental variation increases. As reproduction becomes more concentrated in young individuals, populations begin to fluctuate violently in reaction to environmental fluctuations. When net fecundity distributions characteristic of colonizing species were considered, populations tended to die out completely under some regimes of environmental variation. These computer results make it abundantly clear that fluctuating environmental condition may select for conservative net fecundity distributions with many ages of reproduction rather than all-or-nothing reproduction in which all reproduction occurs in only one or a few reproductive intervals.

TABLE 3-7. Relationship between age of first reproduction, reproductive span, and long-term variation in spawning success of various fishes*

POPULATION	FIRST AGE OF REPRODUCTION (YEARS)	REPRODUCTIVE SPAN (YEARS)	VARIATION IN SPAWNING SUCCESS
Herring, Atlanta-Scandinavian	5-6	18	25x
Herring, North Sea	3-4-5	10	9x
Pacific Sardine	2-3	10	10x
Herring, Baltic	2-3	4	3x
Anchovy, Peru	1	2	2x

*Data from Murphy, 1968. Variation in spawning success is a relative measure. Large values denote great variation. A value of zero would imply constant year to year success.

The reason for this peculiar phenomenon is not immediately obvious. Why should species with low intrinsic rates of increase be more fit than those with earlier reproduction and higher intrinsic rate of increase? The above results indicate that the relationship between a population's age and net fecundity distribution is the key to answering this question. If a population lives in an environment of fluctuating suitability for reproduction, then its age distribution will be in a constant state of flux. However, the proportion of individuals per age class will have some long-term average form. The relationship between this age distribution and strategies of net fecundity distribution may be understood in terms of overall population fitness. One measure of a population's fitness is the geometric average of its realized rates of increase. Define λ, the realized rate of increase as:

$$\lambda = \frac{N_t}{N_{t-1}}$$

Then population size at time $t = x$ must be

$$N_{t=x} = (N_0 \lambda_1) \lambda_2) \lambda_3) \lambda_4) \cdots t = x-1)$$
$$\text{or } N_t = N_0 \prod_{t=0}^{t=x-1} \lambda_t^{1/t}$$

The average of these realized rates of increase is the geometric mean. We already know that:

$$\lambda_t = \frac{N_t}{N_{t+1}} \text{ and } N_t = \sum_{x=1}^{d} (N_{t-x} l_x) m_x$$

where $N_{t-x} l_x$ defines the population's age distribution. Therefore, realized rate of increase depends on the coincidence of $N_{t-x} l_x$ and m_x, or of the age and fecundity distributions (see Table 3-8). A population that

TABLE 3-8. The relationship between age distribution and net fecundity distribution as reflected by λ or r^*

x	l_x	m_x	INDIVIDUALS, POPULATION 1			INDIVIDUALS, POPULATION 2		
			$B_{t-x,0}$	$l_x B_{t-x,0} = N_x$	$N_x m_x$	$B_{t-x,0}$	$l_x B_{t-x,0} = N_x$	$N_x m_x$
0	1	0	40	40	0	—	—	—
1	0.9	0	70	67	0	70	63	0
2	0.7	0.8	70	49	39.2	60	42	0
3	0.6	0.6	60	36	21.6	50	30	33.6
4	0.4	0.5	50	20	10	40	16	18
5	0.3	0.3	45	15	5	30	9	8
6	0.2	0.2	35	17.5	3.5	55	11	3
7	0.1	0.1	50	5	0.5	90	9	2.2
8	0	0	70	0	0	95	0	0.9
					$\Sigma = 79.8$			$\Sigma = 65.7$

$\lambda = 79.8/40$ = 1.995

$r = l_n \lambda = 0.6906$

$\lambda = 65.7/70$ = 0.9385

$r = l_n \lambda = 0.9385$

*Example populations have the same l_x and m_x distributions, but age distributions differ. Even when total young born ($B_{t-x,0}$) are the same, the population with better coincidence of N_x and m_x has a far greater rate of increase than the others in which most individuals are either pre- or postreproductive.

has its age and fecundity distributions more constantly in coincidence will have the highest average rate of increase. Since populations with broad fecundity distributions are more likely to have these distributions coinciding for higher proportions of time than ones with more concentrated reproduction, they may have a higher average realized rate of increase. The effect of the relationship between age and net fecundity distribution and rate of increase depends on the way the environment fluctuates. If the environment is such that the population has an average growth rate of zero, populations with broad net fecundity distribution are superior. Usually, favorable environments favor populations with early reproduction and high intrinsic rate of increase.

Limitations affecting species—net fecundity distribution and k selection

Population geneticists in general feel that selection usually maximizes a population's fitness, and that the value of this fitness reflects the population's innate ability to increase. However, it has been clear to ecologists for some time that the situation is not a simple one. Most species are limited by the carrying capacity, k, of their environment. From the Lotka-Volterra equation, $dN/dt = rN(1 - N/k)$, we have a definition that states that maximization of fitness should involve modification of the way species use or react to the biotic portion of their environment that determines their ultimate numbers. The components of this portion are the k variables—food quantity, available space, and predators—all of which act in a density-dependent manner. Obviously, reaction to food shortage and predation can take the forms of increased efficiency on the one hand, and escape from predators on the other. These treatments will be considered later. However, modifications of the net fecundity distribution are also of some importance to k-selected species.

Much of the work done on k-selected species has been done on birds. From this source one soon realizes

that birds have a limited, almost constant number of young per year. Modifications in their net fecundity distributions are generally of two forms, both of which tend to maximize survivorship of parents and minimize energy wasted on excess young. Most species exhibit parental care of young, which primarily consists of feeding young until they are capable of capturing their own food. Therefore, clutch size in birds (and litter size in mammals) appears to be governed by the maximum number one or a pair of parents can successfully feed and rear.

A secondary set of constraints concerns the cost of offspring to the parents. Too many young might overtax parents, weakening them sufficiently that parental mortality increases. David Lack has found evidence to suggest the existence of this relationship in many species of passerine birds. Other investigators have found that birds with more than their maximum number of young fail to successfully raise *any*. For example, in one species the parent cannot incubate more than two eggs. When the occasional female lays three eggs, all three perish because she incubates two at a time, but never the same two. An egg not incubated cools and dies, so eventually all three eggs perish.

Although insect-type colonizers minimize development time and mean age of reproduction to attain an r high enough to compensate for high death rates of young, adults of k-selected species seem to have sacrificed short development time and peaked reproductive distribution for increased parental survivorship and long-term reproductive security. In a sense, the colonizers can be likened to the financial world's speculators, risking most of their capital on the chance of a fortune; the k-selected species buy bonds hoping for small but steady and, on the average, positive return on investment capital.

In birds clutch size varies with latitude. This variation is thought to reflect either degree of exposure to a harsh environment or amount of food available per parent bird. There is evidence for either interpretation of the reasons for latitudinal variation in clutch size. Clutch size is usually greater in the north than in the south and, in general, falls to lowest levels in tropical regions. Migrating birds and northern forms have large clutches, as might be expected of species in which there is uncertainty of survivorship during the first year because of migration or inclement weather. Tropical birds, which undergo intense interspecific competition (see Chapter 4) have small clutches. Thus neither hypothesis concerning clutch size differences can be ruled out. Perhaps the best evidence for food limitation of clutch size comes from comparing predatory birds with herbivores or insect-eating forms. Large predators, which usually have few young per year, have far less available food per unit area than do herbivores, which usually are not food limited.

Cody has advanced a general theory of clutch size, which takes into account k selection (in the form of predation) and climatic stability (r selection). Cody combines the two selective pressures graphically and determines optimal clutch sizes for any combination of environmental conditions. According to this theory, clutch size should increase as environmental instability increases. Decreased carrying capacity should result in smaller clutch size.

Data on clutch sizes in blackbirds (Icteridae) agree well with theoretical predictions. Number of eggs per clutch increases with latitude from 2 eggs at 10° to about 5 at 50° north or south latitude. Birds from a stable habitat, such as a forest, have fewer eggs per clutch than similar species that occupy less physically stable habitats at the same latitude. Island populations have lower clutch sizes than birds of the same species that live on the more climatically variable mainland. In addition, clutch size of great horned owls, song sparrows, and other birds increases with distance from the climate-stabilizing oceans. Three species of North American warblers, the Cape May, the bay breasted, and the Tennessee, depend largely for food on the spruce budworm. This is an unstable food supply and the warblers that feed on it lay more eggs than other species.

Modification of clutch size is only one way in which a species may adjust its net fecundity distribu-

tion to k selection. According to I. A. McLaren the long life of the chaetognath *Saggita elegans* is such an adaptation. He has collected data on the relationships between body size, temperature, egg production, and mortality rates for several populations. Individuals of arctic populations from a thermally stable but trophically cyclic environment are about twice as large at maximum size as are those from the less thermally stable but trophically more stable waters off Plymouth, England. Animals from arctic populations mature at 2 years, but those from more temperate regions require only 1 year to reach reproductive maturity. The small Plymouth form has five or six spawnings per year, the large arctic form only one every 2 years. Why does the arctic form reproduce less often and grow more slowly than animals from more temperate waters? It would seem that such reproductive behavior must be nonadaptive. The answer to this question comes from consideration of reproductive and mortality data. Fecundity is the same direct function of body size for all populations: egg number per individual = 0.115 length$^{2.46}$. Thus animals that are 10 mm long produce a maximum of 32 eggs each and animals 40 mm long produce 950 eggs each. At reproductive maturity arctic animals are 27 to 38 mm long. Mature Plymouth *S. elegans* range from 10 to 20 mm in length. McLaren finds that the maximum second year mortality rate is 0.616 and notes that mortality during both years of a cohort's life must be the reciprocal of number of births for the 2-year-old mature animals. The resulting net fecundity value for biennial populations is then 1. Number of eggs × the proportion surviving to maturity = $543 \times 1/543$ for a rate of increase of zero. Since mortality and fecundity rates are known, we can calculate that if the arctic population matured in 1 year, its rate of increase would be -0.043. Therefore, extended life is of value to the arctic population.

A compromise must always be made between r and k selection. Most species live in some form of colonizing environment, and most are subject to some degree of food shortage as well. Compromise in life table

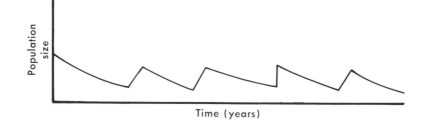

Fig. 3-12. Fluctuations in density of the California condor, stylized. (After Mertz, 1971.)

strategy is often of the form illustrated by data on various sardines and sardine-like fishes discussed earlier.

A final strategy of reproduction is that of relic species, such as the California condor, which may be engaged in a rear guard action in the face of a steadily deteriorating environment. On the average the rate of increase of the condor population appears to be negative. A stylized representation of fluctuations in condor density is shown in Fig. 3-12. The condor's strategy amounts to simply increasing the number of reproductive age classes. The California condor lives at least 50 years, and adults produce one offspring every second year. Even though rate of increase for this species is usually slightly negative, in a good year the presence of old reproducers allows relatively fast population growth because most individuals are old, when chance of reproductive success is greatest. Of course, a glaring disadvantage of such a fecundity distribution is that maximum rate of increase possible must be exceedingly low. This is a specialized strategy indeed! It seems to violate Fisher's fundamental theorem, but, of course, it does not, since if these birds did not have long reproductive lives their chance of survival (and therefore their fitness) would be zero.

The relict seed

A mathematical treatment has been worked out to explain why some plant species produce different kinds of seeds, part of which germinate immediately and part of which require some releaser mechanism to germinate or are simply slower maturing (see Cohen, 1966). The answer derived by Cohen defines an important existence strategy, which is apparently in widespread use by species that live in unpredictable environments. The results indicate that the more unpredictable the environment, the more advantageous it is to have late germinating seeds. If the first batch germinate in an unsuitable environment, the relicts offer an additional chance for reproductive success to the population. This model applies perfectly to the damsel flies mentioned earlier. In the Florida population adults emerge during warm periods throughout the year. Females breeding in the fall produce young that take from 2 to 6 month to mature. Were this not the case, the entire population would risk extinction if a winter breeding interval was too short to enable females to breed and lay eggs.

This relict seed phenomenon is probably a common adaptation in organisms from temporally uncertain environments. It has been demonstrated in desert plants, and probably occurs in the damsel fly. In addition, Murdoch has shown that members of a species of beetle that normally reproduce and die within a single year may delay reproduction until a second year if food level is low during the first year. I have evidence to suggest that single *Drosophila melanogaster* females produce eggs with vastly different development times. This variation appears to be more pronounced in flies captured from uncertain environments than in those from more climatically stable areas. Thus each female produces a proportion of "relict seeds" thereby increasing her ultimate fitness.

Carr reports that the green sea turtles lay many clutches of eggs during a single laying season. The reasons may be both logistic and physiological; no female could possibly lay her yearly quota of more than 1,000 eggs at one time. In fact, work in progress suggests that there is a maximum clutch size of about 200 eggs. If clutches were larger, gas exchange in the subterranean nest would become limiting. In addition, there is reason to believe that the relict seed phenomenon may be operating. The chance of success of any particular nest is small. Some laid too far from the water may become desiccated; others may be destroyed by predators. Thus a female maximizes her reproductive success by spreading her efforts temporally, and as a result, spatially.

A hybrid strategy

As mentioned at the beginning of this section, almost no species follows only one strategy of fecundity distribution. High intrinsic rate of increase and ability to persist in a fluctuating environment often are contradictory. Therefore we may expect to find that populations show evidence of hybrid strategies. Thus in one species of beetle, adults normally reproduce during the first year of life and then die, but if females are faced with food limitation they will reproduce less than the normal amount in their first year, live to a second, and then reproduce (Murdoch). My data (not yet published) indicates that *Drosophila melanogaster* populations differ in net fecundity distribution form depending on the area in which they are found. Populations from climatically stable environments with short growing seasons begin reproduction 1 to 2 days earlier than those from environments in which short-term environmental heterogeneity is great but growing seasons are long. Populations from the first area have shorter reproductive lifetimes and those from the latter type of environment produce a large

fraction of eggs that take a long time to develop (and thus serve as "relict seeds"). Preliminary data indicate that *D. melanogaster* is strongly genetically polymorphic for development time and length of reproductive life (see Chapter 2). At the xanthine dehydrogenase locus (on the third chromosome), homozygote genotypes differ by as much as 2 days in development time. Homozygotes for this section of the third chromosome differ in peak age of reproduction by 3 days! Clearly, this sort of polymorphism can give populations a huge degree of strategic plasticity. The early reproducing genotype will increase in frequency in a growing population, thus hastening rate of growth. When populations decline, they will do so slowly because the late reproducing genotype is selected.

SEX RATIO AND SEX-RELATED DEMOGRAPHY

All of the reproductive strategies mentioned so far have assumed that the female life table is of sole importance. Are such assumptions about the importance of male contributions to population demographics seriously in error?

Only recently has it become obvious that an entire class of species may split the immediate effects of selection unevenly between the sexes (see Giesel, 1972, 1974). These are generally the colonizers, or those that live in a coarse-grained environment in which selection intensities are constantly high. In several species of fruit flies, damsel flies, mice, voles, lizards, and in man, for example, males are either more subject to selection because of their behavior (generally pioneering) or are simply more labile to the selection intensities they share with females.

For example, in *Drosophila melanogaster* males are less able to stand temperature extremes than are females; females of this species are far better at physiological acclimation than are males, and males undergo high mortality when challenged with temperature extremes. In *D. pseudoobscura,* a species that seems to be well buffered to climatic variation through the use of coadapted gene complexes, males and females have equal powers of acclimation and presumably respond similarly to natural selection. However, it has been observed that *D. pseudoobscura* males have poorer survivorship under crowded culture conditions than do females of the species. (See Giesel, 1972, for references.) Could this be a response to *k* selection?

Males often expose themselves to selection more than females do. This is true of voles and mice where young males migrate between populations to a far greater extent than do females. In the damsel fly *Ischnura ramburei*, a species which, at least in Florida, is subject to great environmental heterogeneity and, presumably to changing directions and intensities of selection, males spend more time as prefertile adults than do females, and because of this demographic behavior they have higher death rates. This species is believed to actively track changes in its environment by genetic change. It is likely that this tracking could not occur if environmental effects were evenly shared by members of both sexes, since as this would result in too much loss of fitness. Male assumption of mortality seems also to be of importance to *Myotis austroriparius,* a common cave-dwelling bat. This species is believed by some biologists to be in the process of colonizing Florida and other southern areas from its ancestral range, the limestone caves of the Midwest. In Indiana populations, in which the bats are presumably well-adapted to local conditions, the primary (at birth) and adult sex ratios are nearly equal. There is no excess male mortality in this ancestral range. However, the Florida populations are quite different. Changes in sex ratio indicate conclusively excess male mortality. Investigation of the physiology and behavior of male and female bats reveals that this occurs because in winter the males roost near the mouths of caves rather than deep within the caves as do females. This behavioral difference results in excess mortality in Florida, because the males are exposed to local weather fluctuations. Northern male bats would remain in torpor throughout winter, but in Florida the males wake and must forage for scarce food supplies, and they must avoid predators whenever the temperature rises above 14°C, which happens often. If both males and females lived in the open, winter fitness loss would be excessive. However, when only males are exposed to selection, the population may adapt to new local conditions at almost no cost, or reduction, of rate of increase.

The theory necessary to explain the importance of sexual differences in mortality involves a combination of demographic and population genetic arguments. You recall from Chapter 2 that rate of change of allele frequency

$$\Delta q = \frac{q_0(w - \overline{w})}{\overline{w}} - q_0$$

is dependent on initial gene frequency, the population mean fitness, $\overline{w}$, and the fitness differential (the difference between $\overline{w}$ and w). There is also a lower limit to fitness. If a population's mean fitness is less than 1 on the average, the rate of increase will average less than zero, and the population will not survive. When considering fitness for many populations, we need refer only to females, at least in terms of rate of increase. Males are unimportant to the population's growth rate and survival as long as enough are present at mating time to fertilize all available females, but males may be disproportionately important to the population's evolutionary ability.

During any one generation, change in gene frequency at a locus under selection must be an average of the changes which occur in the male and female members of the population. Change in gene frequency that occurs within a group is inversely proportional to the group's fitness. Thus if males are more sensitive to environmental change and undergo more relative mortality than do females, their gene frequencies will change more rapidly than will those of females during the course of a single generation. The importance of males to gene frequency change is in direct relation to selection intensity among the males. The more males are selected against relative to females, the faster will be gene frequency change for any given reduction in immediate population fitness.

Selective loss in the recent past has been more intense against males than females in man. This phenomenon is general at least among animals. Indeed, it is difficult to think of any animal in which some form of sexual difference in mortality does not exist.

Males are usually more labile to selection than females. This is of value to females and to the entire population. Other differences between males and females have also evolved in order to take advantage of the disparity. Male and female development times often differ. In most animal species males require more time to reach reproductive maturity than females. The advantage of this is twofold. First, if we assume that selection intensity can be measured per unit time giving

$$S = 1 - \mathop{\pi}_{t=1}^{t=d}(1 - S_t)$$

for total selection before reproduction, increased development time of males will allow them more time to bear the brunt of selection. When random mating is assumed, this results in faster change of allele frequency and more efficient population evolution. This effect of late male development is accentuated in iteroparous populations. Only then is late male reproduction compensated by making males of the oldest age classes more important in terms of their contribution to the next generation's allele and genotype frequencies.

The second aspect of development time differences is that females are left free to reproduce early and thus to optimize the population's rate of increase. In rhesus monkeys, some mice, voles, and perhaps other animals, females often produce female offspring first and then produce males. A model can be made by introducing a second term, s_x, to the standard equation of iteroparous population growth. When s_x stands for the proportion of offspring produced by a female aged x that are female, the following defines the rate of increase:

$$I = \sum_{x=1}^{d} l_x m_x s_x e^{-rx}$$

Since the solution of this equation for r is not obvious, we can look at this case as we have treated more simple ones. Remember that:

$$r = \frac{\ln R_0}{T}$$

to a reasonable approximation, where R_0 is net reproduction of female offspring per female life time and T is the average age at which the population's females give birth. With the addition of s_x, R_0 becomes:

$$\sum_{x=1}^{d} l_x m_x s_x \quad \text{and} \quad T \text{ is} \quad \frac{\sum_{x=1}^{d} l_x m_x s_x x}{\sum_{x=1}^{d} l_x m_x s_x}$$

If s_x is large for young females and then declines (if young females have mostly female young and older mothers give birth primarily to males), then female generation time is reduced and R_0 is increased. Thus r_m can become larger. This type of modification may be of primary importance in polygynous species that live in harsh environments and in which females are unable to further increase reproduction by other means.

Only shallow slopes of the s_x function are required to yield large increases in intrinsic rate of increase. However, female rhesus monkeys produce an average of 62% female offspring until age 5 years and 32% females at ages 6 and 7 years. Under the terms of this model, females are expected to increase in proportion in growing populations and decrease in declining ones. The increase in growing populations simply causes rate of increase to grow as well. But what about the decrease during periods of population decline? As females decrease, proportion of males (in early age classes at first) increases. If males are the selection-prone sex, their increased abundance in early age classes should hasten the population's adjustment to whatever factor caused it to decline. The conclusion can be derived mathematically, but no solid evidence indicates its importance as an existence strategy.

Mating success

Populations live in clusters in patchy or heterogeneous environments. One explanation for the patchy distribution of animals is that the environment provides only small and isolated areas suitable for the persistence of particular species. It is probably at least as important to consider the effect of aggregation on population dynamics. Recent work has indicated that in many species populations may be clumped as a result of adaptations leading to certainty of mating success. For example, it can be calculated that if the sex ratio characteristic of a population is 1 : 1, a female searching at random among other individuals of her own species would have to make six contacts before contact with a potential mate becomes reasonably certain. If this same female were searching for a mate in a homogeneous enviornment in which members of the population are dispersed at random, she would have a far more serious problem. This is illustrated schematically in Fig. 3-13. Instead of having to look simply for a male within a dense population of males and females, she must search through a series of areas, each of which may or may not contain a suitable mate. This

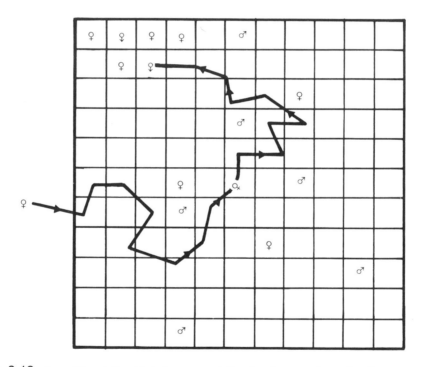

Fig. 3-13. A possible relationship between population density, sex ratio, and mating success. Any female entering the habitat must search through it until she finds a mate. If her "waiting time" until success is long, her contribution to population increase will be small, because her generation time might be longer than that of other females. Her fitness is given by $w = \sum\limits^{x=d} l_x m_x e^{-r(x+t)}$

where t is the excess time she takes to mate. The model suggests why most species have evolved sexual dimorphism and other mechanisms to increase speed of mating.

problem translates directly to rate of increase. As far as r_m is concerned, the most important derivative of the life table is the generation interval, T. This is easily thought of as the age at which the population's females give birth. In a population of a semelparous species, in which there is only one age of reproduction, T is the age at giving birth. Ideally this should be the same as the age of reproductive maturity. However, for the searching females just hypothesized, T is extended by the time elapsed between the time females first become reproductively mature and the time at which a mate is found. This elapsed time should be some simple function of the number of trials a female must perform before locating a mate. Whatever form the function takes, it should be clear that by increasing T, random dispersal reduces a population's rate of increase.

Various mating behaviors serve to reduce the element of chance when finding a mate. There is probably no bisexual species that is devoid of sex attractants, derived sexual behavior, sexual dimorphism, or some other modification to render this search nonrandom. Such modifications are extremely well known. However, when considering the second part of the problem faced by our anxious female, are there in fact modifications of population structure that act to reduce environmental randomness? The answer, of course, is yes; most species have clumped distributions. Fish travel in schools, birds in flocks, wolves in packs, etc. Among insects, male dragonflies and damsel flies occur often along edges of streams and ponds where wandering females can easily find them. In addition, the males of most species of odonates are bright colored and territorial. Both of these characteristics serve to attract females once they have located the centers of male population density.

Population density is of critical importance in population dynamics. If populations are too small or diffuse, reproductive success may be small or nonexistent. On the other hand, most populations are faced with some sort of upper limit to population size. The approach of a population to the upper limit often trig-

gers dispersal or random out-migration from existing populations.

DISPERSION AND DISPERSAL
Genetic and regulatory mechanisms

As noted in Chapter 2 migration is one of the more important genetic processes of natural populations. This is particularly true if migrants are among small subpopulations or colonies that would otherwise be subject to loss of genetic variability resulting from genetic drift.

Dispersal is a mechanism of interpopulation migration and seems to have been selected for in many populations. The process is triggered by a variety of factors, but perhaps the most important of these are changes in weather and increased population density.

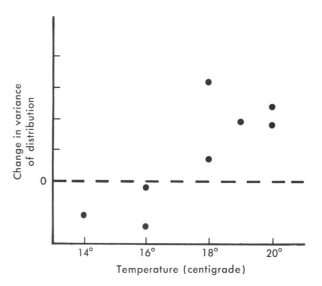

Fig. 3-14. Relation between temperature and dispersal in *Drosophila pseudoobscura*. Dispersal is high at temperatures above 17° C but "negative" at lower temperatures, perhaps due to inactivity and not to a real shrinkage of the distribution. Data points are changes in variance of a distribution of flies released at a single point. Points above dashed line represent positive dispersal of flies. (Data from Dobzhansky and Wright.)

The influence of weather on dispersal has been shown in a large number of studies.

Fig. 3-14 illustrates a relationship between dispersal of *Drosophila pseudoobscura* and temperature. As temperature increases, so must the mobility of insects whose metabolic rates are a direct function of temperature. It is not as easy to decide why negative rates of dispersal are shown below 15° C. Perhaps the animals simply become difficult to find on a cold day. This would give the appearance of a contracting population.

Dispersal is a mechanism by which densities of parent populations are decreased and new areas are colonized. Thus weather-responsive dispersal is a process which we might expect to be selected for in species with high rates of increase.

The western tent caterpillar, *Malacosoma pluviale,* is genetically polymorphic for dispersal tendency to a startling degree. Populations of this species consist of four phenotypes. Individuals of types 2, 3, and 4 are inactive as larvae and metamorphose into sedentary adults. Those of type 1 are active during the larval stages and are highly migratory as adults. Types 2, 3, and 4 form a series of decreasing activity. An adult female of any type can produce larvae of all four types, but type 1 females yield large proportions of type 1 larvae, etc. The effects of this polymorphism on population dynamics are adaptive because the species occupies an unstable environment. When weather conditions are favorable, type 1 females leave parent colonies to colonize marginal habitats. Many of their progeny are also of type 1, and they also colonize new habitats. This results in a rapid range expansion for the population. The moths left behind in old colonies are eventually mostly of types 2, 3, and 4. Because these phenotypes are inactive and have low viability, old colonies may become extinct unless they receive influxes of type 1 females from other colonies. When environmental conditions deteriorate, most populations of migrants and all populations of inactive moths die out. The only populations that survive such a collapse consist of migrants which happen to be in more favorable microhabitats than others. In this way the species continually seeks out and colonizes new habitat patches. Thus the species persists because it has evolved a mechanism to avoid heavy density-dependent selection (which would often be significant if the population remained in one small area); it adapts through constant migration (and founder effect?) to a harsh, fluctuating environment.

Weather and dispersal are related in yet another way much more directly to probability of species persistence. Lewontin and Prout's work with *Drosophila pseudoobscura* illustrates this nicely (see Chapter 2). This species consists of selectively different local populations, within and among which genetic variations and adaptability are probably kept at high levels via dispersal and population migration. Dispersal of many marine intertidal invertebrates is also directly related to environmental harshness. These organisms are exposed daily to both the relatively constant marine environment and to the far more variable environment when the tides recede. Thus selection is intense among these forms and, depending on latitude and local conditions, chance of local population extinction is variable but generally quite high. There are two general forms of reproduction found in these animals. Mating may occur resulting in young which are attached directly to the substrate near the parent organism. Dispersal of young is obviously low in these forms. At the other end of the scale are forms in which spawning occurs. Parental organisms release eggs and sperm into the water; these unite to form free-floating zygotes, which are swept offshore by the receding tide. These are free-floating or swimming plankton. The planktonic larvae of some invertebrates remain at sea for long periods and travel for great distances before settling. Others exist solely on their own yolk masses and are planktonic for short times only.

Dispersal varies from species to species. It seems to depend to a great extent on probabilities of local population extinction. For example, snails that have no operculum (a structure that allows the animal to form a tight seal and isolate its visceral mass from desiccation or fresh water), have large numbers of planktonic, long-lived larvae. Opercular forms often deposit their eggs in sticky masses on the substrate and the young are either planktonic for only a short time or not at all. Apparently, amount of dispersal is directly correlated with the harshness of the adult's environment.

The chance of the death of the young is also an important part of the mortality-dispersal equation. In arctic populations of marine invertebrates, for which resources of larval stages are unpredictable and generally limiting, the planktonic larval stages are either short-lived larvae or there is no planktonic stage at all. Their more temperate zone congeners have longer dispersal stages, perhaps because larval survival is more certain.

For many species dispersal is related to population density or, more correctly, to the nearness of population size to carrying capacity, as denoted most simply by the logistic equation of population growth. There is much literature on this subject; the most appealing work has been done with mice and voles. In these species proportion of dispersants increases directly with nearness of N to k. Dispersants are most commonly young males. If you think about this, the evolutionary reason becomes clear. Males use nearly as much resource (food, space) as do females, but at least in voles and mice one male is sufficient to fertilize many females, so males are expendable. At the same time, males are the weaker sex in terms of ability to withstand inclement weather and competition. Thus the loss of young males costs the population nothing in terms of growth. At the same time, should any of the migrants, which suffer high intensities of selective death during migration, immigrate to a new population and become part of its breeding pool, they will contribute additional genetic variability and fitness to their adopted populations.

Dispersal and population density

The logistic equation implies that stress on members of a population increases with increased population density. This effect could be alleviated by out-migration from centers of populations density. Such migration has been well-demonstrated in muskrats. Errington observed that older members of a colony banish their young during periods of high population density. The advantage of this population control to the parent population is obvious, but its effect on individual fitness needs some discussion. At first, the fitness of the parents of these waifs appears low; most of the migrant muskrats fell prey to carnivores. What then is the advantage in producing excess young to be turned out to fend for themselves? Advantage to the parents of migrators might accrue because habitat patches differ in their carrying capacity in both time and space. If a patch has sufficient resource to support a large population at one time but a much reduced capacity shortly thereafter, and if nearby patches also fluctuate in carrying capacity but asynchronously, then there might be some advantage to producing young with a propensity toward inter-patch migration.

As calculations done by Gadgil show, interpopulation migration can be a fine-tuned mechanism for maximizing a species density and fitness in a resource-limited, patchy environment. With optimum migration, no population need ever exceed its carrying capacity. Investigation of density-mediated dispersal and its possible effects on community structures and dynamics is pursued in Chapters 4 and 5.

CYCLES OF POPULATION DENSITY

Ecological literature and folklore are replete with examples of species cycling in population density. Some of these changes in density are relatively easy to explain on the basis of what we already know about rate of population increase. For example, most insects are seasonal in abundance, with low ebbs of population at certain seasons and peaks at other times when growing conditions are more favorable. In Florida the damsel fly, *Ischnura ramburei,* shows peaks of density in June and October. These peaks are supplemented by small bursts of activity in late fall and winter during warm periods. The winter of 1972 was unusually warm, allowing more reproduction than usual. As a result, the population the following summer was very dense. Comparison of 1972 with 1971 suggests that in addition to the seasonal cycles of abundance in this animal, there are irregular changes in abundance, which can be ascribed to yearly changes in weather. The same might be said for the numerical fluctuations of many other insect species.

Although some changes in population density or number are easy to analyze, others are far more difficult to explain. Many cannot be satisfactorily explained, and others can be at least partially explained but only in terms of population and community regulation and stability (see Chapters 4 and 5). In this section we will discuss only those cases that seem to be relatively self-contained and that do not require analysis of community structure for explanation.

Watt cites interesting correlations of density fluctuations with periods of abnormal weather. Fig. 3-15 shows the density cycle of the Canada lynx, for which trapping information is available over a period of 80 years. This cycle is more or less regular with significant low densities appearing in 1889 and 1918. Watt has attempted to correlate these changes in lynx density with data collected on temperature variation for the years 1823 to 1916 and finds that weather variation and lynx density variations are significantly cor-

related. These data were taken from a broad geographic area, which tends to support the importance of weather fluctuations as causative agents; if some local disturbance were responsible we would expect to find local deviations from global density.

Calculations show that major minima of lynx density (1889, 1918) are highly correlated with eruptions of the major volcanoes Krakotoa and Katmai in 1883 and 1912, respectively. Lynx density was also at a nadir in 1842, which was 7 years after the eruption of Cosequinas. Major volcanic eruptions cause major depression in global temperature for 5 subsequent years and somewhat lesser effects for an additional 5 years. Statistical analyses suggest that colder weather than normal results in low lynx density.

Canada lynx depend for much of their livelihood on the snowshoe hare, and data suggest that changes in population density of this animal are also related to weather patterns via plant growth and density. Therefore, following a series of cold, wet years there should be a decline in rabbit density because of starvation, which would lead in turn to underfed and less reproductive lynx with poor juvenile survival. A few years hence a nadir of lynx density should occur. (Watt notes certain relationships between lynx density cycles and

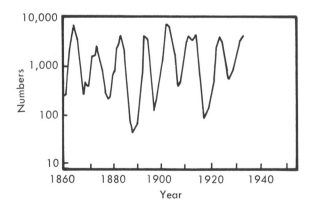

Fig. 3-15. Fluctuations in lynx density in the MacKenzie River district of Canada. (Data from Elton and Nicholson, 1942. After Watt, 1972.)

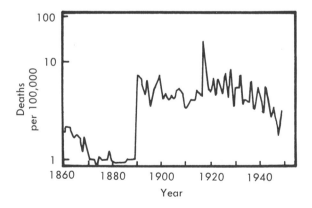

Fig. 3-16. Influenza death rate per 100,000 individuals in England and Wales. (Data from Andrewes, 1953, and Deutschman, 1953. After Watt, 1972.)

influenza epidemics also. Peak years for influenza in England and Wales correspond to nadir years for the lynx.) (See Fig. 3-16.)

Influenza epidemics in man might well be related to cold and wet weather. One study shows quite conclusively that incidence of influenza increases with decreasing daily minimum air temperature. Another study shows that during influenza epidemics those individuals who have outdoor occupations are more heavily infected than are those with indoor jobs. Therefore, outbreaks of influenza are most common in cold weather, but only because the host is more susceptible to the disease then. Just as lynx are predators on hares, influenza virus is a "predator" of man and increases with increased "prey" density, in this case availability of susceptible hosts.

Other epidemics that show definite relationships with weather conditions are Thai fever and plague. Thai fever incidence is associated with periods of heavy rainfall, which increases the population density of the mosquitoes responsible for host-to-host transfer of the infectious organism. Plague epidemics seem to be related to humidity as well. Dry weather lowers survival of rat fleas, which are necessary for plague transfer.

In general, population densities of insects in Canada seem to be related to temperature. The relationship is probably both direct and indirect; an insect species' rate of increase is directly related to temperature, and the resources of insect population are affected by temperature-induced changes in their consumers' density.

There are still problems of population stability that puzzle investigators, such as the problem posed by lemmings, voles, and mice. Most people are probably at least partially familiar with the lemming story. These rat-sized mice go through tremendous cycles of population density, each peak of which is culminated in the now-famous march to the sea and, apparently, mass suicide by drowning. Perhaps it is not too difficult to explain why lemming populations grow to peaks. These animals have short generation times and

a high rate of increase, and if no resource is limiting, they will attain large population densities rapidly. Eventually, however, densities apparently get too high and the animals abandon their normally secretive ways to run about openly on the tundra. According to one writer, when this occurs they become ready prey for hawks, foxes, snowy owls, and other predators, and eventually they begin their march. Several authors have suggested that this behavior is the result of social stress caused by high density and increased interindividual contact. No one knows for sure.

Field voles, which are distantly related to lemmings, appear to go through fairly regular, 3-year density cycles. The work of Krebs and others (see Gaines and Krebs) can offer no direct cause for this cycling, although at least one explanation has been tendered by Chitty. He suggests that a particular genotype for aggressiveness is favored at low population densities, since it has as a side effect increased activity by males and perhaps increased chance of male-female encounter. Being favored by natural selection, such a genotype would tend to become more frequent with increased population size. But Chitty suggests that at some critical population density, the fitness of aggressive individuals switches from good to poor because of increased male-male and female-female contact and resultant fighting. Chitty then reasons that the population's death rate should increase and its birth rate decrease because of prevalent stress conditions. Thus the population would be expected to begin a decline phase and to decline until some lower limit to population density is reached whereupon aggression would again be lowered. Although many population biologists have worked to corroborate Chitty's hypothesis, none has succeeded conclusively.

However, some of the data collected by Krebs and co-workers may support Chitty's model. Krebs has spent several years following changes in allele frequency at the transferrin locus in *Microtus ochrogaster* and *M. pennsylvanicus*. This work demonstrates that the alleles Tf^E and Tf^F go through frequency cycles that correspond to cycles of population density. Furthermore, heterosis is demonstrated by their results. These data could support Chitty's hypothesis. However, they also fit a model of age-dependent selection suggested by Charlesworth and Giesel (1972). (See Chapter 2.)

Many cycles of population size may be explained in terms of changes in the quality of both abiotic and biotic enviornmental components. In fact, these changes underly all population fluctuations. However, it has recently become clear that the form of a species' fecundity distribution, when driven by extrinsic environmental fluctuation, can act to magnify population fluctuations. The explanation is closely related to the fecundity strategies previously discussed. There are three basic types of fecundity distributions. Of these, the fecundity distributions exemplified by both fish (Murphy, 1968) and condors (Mertz, 1971) tend to

confer numerical stability. The reason for this is clear. In both cases there are many subsequent ages of reproduction. Changes in environmental quality capable of causing either increase or decrease in population size tend to be damped by passage of either large or small age classes through the life table. Large lags in change in population size are built into such a system with the result that population size is relatively stable. In addition, species with attenuated reproductive life generally have low intrinsic rates of natural increase, which act to damp any pulses of population size. The fecundity distribution of the colonizing species is quite different in this regard. In this case there are only a few ages at which reproduction occurs, the generation interval is short, and net reproduction per female is high giving a net result of high intrinsic rates of increase. Of more importance is the fact that such a fecundity distribution consists of only one really productive age interval followed by a series of ages of decreasing reproductive potential.

Fig. 3-17 shows the net fecundity distribution of a typical colonizing species. This overlies a set of age distributions represented by curves. The lines represent a single, extremely good year for reproduction (T_1) and the progress of the resulting age class, or cohort, over the life table distribution $(T_2, T_3, \ldots T_8)$. When the large year class is age 3, there will be another year of unusually high reproduction, and this will be followed 3 years later by still another population peak. The interval between these peaks will be years of increasing population size. If a single poor reproductive year immediately follows the good one, there is a peak year for the population followed immediately by a year of low population size. In further illustration of this point, two computer simulations of population existing in a randomly varying environment are shown in Fig. 3-18. Population B (dashed line) has a conservative fecundity distribution, whereas population A is a colonizing species. Population B exhibits rather violent cycles of density, but those characteristic of A are far less extreme.

The net fecundity distribution of snowshoe rabbits

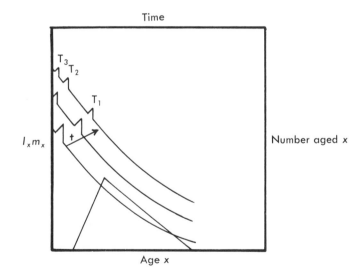

Fig. 3-17. Progress of pulses in age distribution caused by some extraneous factor. Pulses are large at first, then damp as time passes. Successive pulses (T_1, T_2, T_3) become smaller because of the averaging effect of age distribution. At T_1 all ages are affected by the perturbing factor. At T_2 and T_3 pulses are the result of the T_1 birth class passing over the m_x distribution.

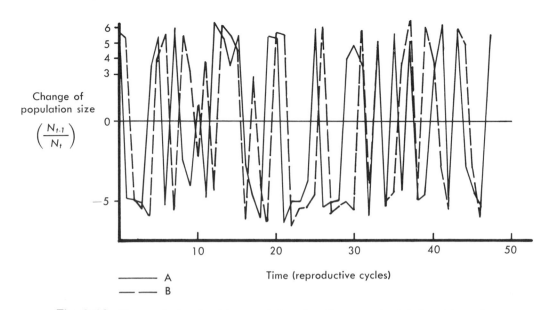

Change of
population size

$$\left(\frac{N_{t-1}}{N_t}\right)$$

Time (reproductive cycles)

——— A
— — — B

Fig. 3-18. Changes in density of two populations from the same simulated random environment. Females of population *A* have fewer ages of reproduction than those of *B*.

is rather similar to that of population B. The lynx population discussed earlier has a single year of low population size followed by periodic increases in number. Lynx cycles may be the result of occasional perturbations of population densities of their food supply (the snowshoe hare), these being propagated through successive generations as a result of the form of the hare's net fecundity distribution.

BIBLIOGRAPHY

Bajema, C. J. 1971. Natural selection in human populations. John Wiley & Sons, Inc., New York.

Carr, A. Personal communication.

Chitty, D. 1960. Population processes in the vole and their relevance to general theory. Canad. J. Zool. **38:**99-113.

Cody, M. L. 1966. A general theory of clutch size. Evolution **20:**174-184.

Cohen, D. 1966. Optimizing reproduction in a randomly varying environment. J. Theoret. Biol. **12:**119-129.

Cole, L. C. 1954. The population consequences of life history phenomena. Quart. Rev. Biol. **29:**103-137.

Cox, G. W. 1969. Readings in conservation ecology. Appleton-Century-Crofts, New York.

Dunbar, M. J. 1960. The evolution of stability in marine environments. Natural selection at the level of the ecosystem. Amer. Naturalist **94:**129-136.

Errington, P. L. 1946. Predation and vertebrate populations. Quart. Rev. Biol. **21:**144-177, 221, 245.

Evans, F. C., and F. E. Smith. 1952. The intrinsic rate of natural increase for the human louse, *Pediculus humanus* L. Amer. Naturalist **86:**299-310.

Gadgil, M., and O. T. Solbrig. 1972. The concept of r-and k-selection: Evidence from wild flowers and some theoretical considerations. Amer. Naturalist **106:**14-32.

Gaines, M. S., and C. J. Krebs. 1971. Genetic changes in fluctuating vole populations. Evolution **25:**702-723.

Giesel, J. T. 1972. Sex ratio, rate of evolution, and environmental heterogeneity. Amer. Naturalist **106:**381-387.

Geisel, J. T. 1974. The relative fitness of populations polymorphic and monomorphic for net fecundity distribution. Amer. Naturalist **107.**

Giesel, J. T. 1974. On the evolution of hermaphroditism and separate sexes. In press.

Hall, D. J. 1969. An experimental approach to the dynamics of a natural population of *Daphnia galeata mendotae*. Ecology **45**:94-112.

Hamilton, A. G. 1950. Trans Roy. Entomol. Soc. London, **101**:1-58.

Lack, D. 1967. The natural regulation of animal numbers. Oxford University Press, London.

Lewontin, R. C. 1965. Selection for colonizing ability. In H. G. Baker and G. L. Stebbins, editors. Genetics of colonizing species. Academic Press, Inc., New York.

MacArthur, R. H., and E. O. Wilson. 1967. The theory of island biogeography. Princeton University Press, Princeton, N. J.

McLaren, I. A. 1966. Adaptive significance of large size and long life of the chaetognath *Sagitta elegans* in the Arctic. Ecology **47**:852-855.

Mertz, D. B. 1971. The mathematical demography of the California condor population. Amer. Naturalist **105**:437-455.

Murdoch, W. W. 1966. Population stability and life history phenomena. Amer. Naturalist **100**:5-11.

Murphy, G. I. 1968. Pattern in life history and the environment. Amer. Naturalist **102**:391-403.

Taber, R. D., and R. F. Dasmann. 1957. The dynamics of three natural populations of the deer *Odocoileus hemionus columbianus*. Ecology **38**:233-246.

Watt, K. E. F. 1972. Principles of environmental science. McGraw-Hill Book Co., New York.

Wellington, W. G. 1969. Qualitative changes in populations in unstable environments. *Canad. Entomol.* **96**:436-451.

4 POPULATION REGULATION

The next step in our treatment of population dynamics is the investigation of why natural populations seem to have remarkably constant densities. We will find that this is largely the result of a set of environmental variables that act on populations in a density-dependent way. Regulation relative to resources will be discussed using several species case histories. Then the effects of interspecific competition, predation, and parasitism will be treated.

FOOD SUPPLY

Changes in population structure and reproduction occur as a result of changes in food supply. Our first model population will be *Lymnaea elodes,* a snail found throughout the Midwest, which has been studied experimentally by Eisenberg in Michigan. *L. elodes* is a common inhabitant of small semipermanent ponds in which water levels vary during the course of a year. Water volume and pond bottom area are greatest in early spring following snow melt; through the summer months there is a more or less steady decline in pond size, culminating in near drying in early fall. Such changes in pond size are cyclical and roughly predictable from year to year. It is reasonable to assume that as ponds shrink, the snail populations will be compressed into more densely populated areas and that food, or perhaps space, will become progressively more limiting.

Eisenberg's study was designed to test the evolutionary results of such cyclical changes in crowding and resource abundance. He began by censusing several populations to determine naturally occurring densities. Following this initial census period, an experimental pond was chosen and subdivided by wire mesh enclosures into segments of equal bottom area. Snails were introduced into these cages in two treatments at three densities, low, normal, and high. For one treatment cages were stocked only with snails of reproductive age. For the other, reproductively immature animals were used. At regular intervals these experimental populations were censused and size and number of egg masses (birth rate), death rate, and growth rate were determined.

Eisenberg's results are presented in Table 4-1. Several generalities are immediately apparent. First, there are no differences in death rates or in individual growth rates among the three density treatments of snails of reproductive age. Birth rates, however, differ significantly. Low density treatments have higher than normal birth rates, but high-density populations exhibit lower than normal number and size of egg masses. Among the reproductively immature, death rate was higher than normal in dense populations and lower than normal in less dense treatments.

Individual rates of growth followed an exactly opposite pattern. Rates were low in crowded populations and high in the less densely populated cages. It is apparent that the snails' reaction to crowding involves all three parameters that determine *r:* birth rate, death rate, and development time or generation interval. Less dense populations respond to their circumstances

TABLE 4-1. Response of *Lymnaea elodes*, a pond snail, to food supply*

	ADULT SNAILS (REPRODUCTIVE AGES)		
		RESPONSE	
CONDITION	l_x	m_x	GROWTH RATE
Normal food	NC†	NC	NC
Low food	NC	Decreased	NC
High food	NC	Increased	NC

	YOUNG AND POST REPRODUCTIVE SNAILS		
		RESPONSE	
CONDITION	l_x	m_x	GROWTH RATE
Normal food	NC	NC	NC
Low food	Decreased	NC	Decreased
High food	Increased	NC	Increased

*Data from Eisenberg.
†No change.

with an augmented rate of increase, whereas crowded populations have reduced rates of increase. Furthermore, it is possible to calculate that the rate of increase characteristic of the three population types will result in their assuming population sizes commensurate with the carrying capacities of their environments, and that they will do so within a few generations.

Food is the limiting factor in this experiment. Addition of supplementary food to the crowded populations resulted in reduction of death rate and development time and augmentation of birth rate. Apparently this species has evolved the ability to adjust to fluctuating food supplies efficiently. In a sense *Lymnaea elodes* "tracks" fluctuations in food supply; but unlike the genetic trackers mentioned in Chapter 2, it does so by modifying its life table distribution, not its genetic composition. Such a pattern of resource utilization makes good sense for this species. The resource is periodic and predictable, and by continually adjusting population size and reproductive effort in response to changes in resource quantity, the species is able to

remain at or very near the carrying capacity of its environment. Thus very little resource is wasted. This is a fine example of the result of *k* selection.

The sheep blowfly, *Lucilia cuprina,* provides an evolutionary and ecological contrast to *Lymnaea.* Adults of this species do not feed; all feeding is accomplished by larvae which are deposited as eggs in open wounds on live sheep. Such wounds are rather uncommon except during lambing time when newborn lambs have open umbilical wounds. At other times female blowflies must find sheep with accidentally acquired wounds. Obviously, food as such is not a critical variable for *Lucilia;* once eggs are laid the emerging larvae are guaranteed unlimited sustenance. What is of critical adaptive importance to this species is availability of egg laying sites.

Since these sites are uncommon and quite unpredictable, the species should have evolved to take maximum advantage of available reproductive opportunities. In a now-famous experiment A. J. Nicholson determined the relationship of blowfly population dynamics to food supply. Flies were raised in screen enclosures, and each enclosure was provided with a supply of fresh liver at one-generation intervals. Experimental populations were initiated with small numbers of adult flies. The results of one of these experiments are shown in Fig. 4-1. The populations of blowflies follow a regular cycle of population density; small numbers of adults produce large numbers of stunted, reproductively incompetent young, the offspring of which are small numbers of well-fed, robust adults. The population density fluctuates violently in relation to carrying capacity; in alternate generations the resource is either underused and wasted or overused. Obviously, these flies have not evolved in response to *k* selection. Rather, evolution has been for maximal utilization of occasional reproductive opportunity, a clear case of *r* selection.

In fact, it is of advantage for female blowflies to deposit their eggs in already infested wounds; large numbers of larvae increase each other's ability to feed through increased rate of breakdown of host tissue.

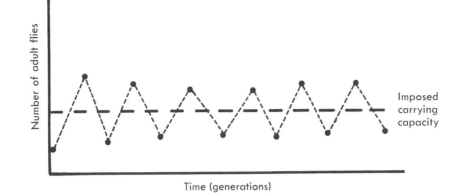

Fig. 4-1. Schematic representation of response of laboratory populations of the sheep blowfly *Lucilia cuprina* to a constant food level. (The actual data were far more irregular than this representation.)

Although Nicholson's experiments may be ecologically unrealistic, they serve to elucidate factors that have contributed to evolution of population dynamics of the blowfly.

SPACE AND TERRITORIALITY

In the ecological literature there has been considerable polemic about the role of limiting resources. Most ecologists now agree on the broad outlines of the issue. Although food may be a limiting factor for herbivores, most are viewed as being primarily limited by space or one of its derivatives (e.g. predation, emigration, and density-related aggression). In contrast, predators are commonly thought to be primarily food limited. To investigate how space is a primary limiting factor we will examine a study of the Scotch red grouse done by Watson and co-workers.

The red grouse occurs in the uplands of Scotland and exhibits strong territorial behavior. Although both food and space are limited in the environment of this species, territoriality is of primary importance to regulation of population size. Territorial behavior is commonly exhibited among birds, mammals, insects, and lizards, usually as a part of mating behavior; it seems to serve the dual function of guaranteeing mated pairs sufficient food for their young and assuring that mating will occur. In the red grouse territoriality serves a quite different function. Most animals are territorial only at the time of mating and when young are being reared. However, red grouse males take territories in early fall, attract one or sometimes two females to share them, and then hold the territories during the winter months. Territorial males and their prospective mates spend the winter months in shrubland where food is abundant and where there is adequate shelter from inclement weather and predators. Males that fail in competition for desirable territories and habitats in the fall are forced to winter on the open moors where food is relatively scarce and where predation and disease result in high mortality. Should a territory become vacant between October and early March, nonterritorial birds compete for and eventually fill it. However, vacant territories are not filled after early March.

It should already be obvious that territoriality tends to assure survival of a fixed number of the population from year to year, since territorial males and their mates suffer low mortality rates. However, in this case regulation of population size is even more indirect because only territorial males mate. Such behavior has a number of interesting consequences: (1) The breeding population (and therefore the entire population) is buffered against fluctuations in size. Since breeding area remains constant, only a certain number of birds can produce young each year; all of the remaining individuals are nonreproductive. Should reproduction fail one year resulting in a decrease of total population size, adjustment is possible since a larger proportion of males take territories the following fall. (2) Territoriality, or competition for space, guards against the species overextending its food resource; the population remains within the carrying capacity of its environment. (3) Territoriality assures that males have been subjected to selection for feeding efficiency (k selection), ability to avoid predators, and ability to withstand inclement weather before their first mating. (4) Since young birds are usually unable to claim territories during their first potential breeding season, they become part of a wandering population. Those that survive the harsh conditions typical of the moors are just as likely to join some adjacent breeding population as they are to become a reproductive part of their original population. Thus exchange of genetic information among populations is assured with attendant reduced loss of intrapopulation genetic variability.

Competition for space serves many purposes, all of which tend to increase a species' long-term fitness. Muskrats also regulate population size below their environment's trophic carrying capacity. They do so by ejecting extra young from home populations. Many of these homeless migrants become food for predators such as mink and foxes, but those that manage to survive either join new populations or found entirely new colonies.

Since space often serves as a factor to limit population size and predators superimpose limitations of space, it follows that predation must be a density regulatory factor in ecological systems. This is certainly true of the red grouse and the muskrats. (Rather than discussing predation here, we will defer it to the end of this chapter and to Chapter 6.) In general, the effects of predation and space limitation are similar. Both act to regulate population size below trophic carrying capacity and both select for generally more fit prey populations.

SOCIAL FACTORS

There is another process that may tend to act in a regulatory fashion in natural populations. Our only evidence pertaining to this phenomenon, which for lack of a better term is called "social aberration," comes from the study of laboratory populations. Populations of mice, when kept in cages in which food and nesting space are supplied in superabundance, often suffer high mortality and cessation of reproduction. Investigation has revealed that dominant males become despotic regarding food and water supplies and attack other individuals indiscriminately. In addition, normal mating behavior becomes subverted, and females with young tend to lose their maternal behavior. As a result, nests are deserted and young either starve or are trampled. The fighting that occurrs in these cages increases the number of wounds, and these become infected.

Similar results occur in field populations. Krebs has found that aggression increases in intensity in large and near peak populations of the field voles, *Microtus ochrogaster* and *M. pennsylvanicus*. In a study of wild rats in Baltimore, Maryland, Christian and Davis found that populations are rather strongly isolated to city blocks. As isolate size grows and crowding increases, adrenal weight increases above normal. This also happens in house mice, in which increases in adrenal weight correlate well with delayed onset of puberty. Such endocrine changes also result in slowed growth rate, inadequate lactation, and increased intrauterine death rates. Further stress in such animals is often lethal. Although endocrine changes have been found in several other species, such as woodchucks and rabbits, they are by no means ubiquitous.

Other studies have shown no correlation between

population size and endocrine weight. Emlen (1973) notes that such change in endocrine activity is probably not an evolved means of population control. It is far more likely that it is a reflection of overexpression of traits evolved for normal-sized populations. The social stress of overcrowding causes subversion of the purposes of aggression and territoriality, which are normally of adaptive value.

The factors that limit herbivore populations may be arranged in a hierarchy. The order of this hierarchy depends upon the loss in fitness incurred by populations as a result of exceeding the bounds of the limiting factor. Exceeding trophic carrying capacity must have serious detrimental results. Overgrazing very often causes long-range changes in vegetation structure and composition, reducing the carrying capacity of an area. Therefore, it is reasonable to suggest that of all limiting factors, food supply should occupy the highest, most protected position. It is also reasonable to place predation in the next position, if only because it seldom becomes an important regulator of a population until the population has exceeded the bounds imposed by space. It is also true that disease, a form of predation, usually becomes epidemic only under crowded conditions. Finally, at the base of the hierarchy, is space, the limiting factor that must be exceeded before either predation or food shortage can become important to a population. It is difficult to assign a position to social density aberrations. They seem to become important only when populations have managed to exceed other lesser bounds set by their environments, and yet in some sense they may act ultimately to protect food supply, at least on a long-term basis.

ADAPTATIONS TO RESOURCE SHORTAGE

The preceding discussion has stressed that many species populations are sometimes limited by resource shortage. There is, however, an additional facet of the resource limitation problem: How do species populations adapt to resource shortages and what trade-offs are the result of such adaptation? Other questions arise from observations of natural populations. Among these are: (1) If proximity is sufficient to cause aberrant behavior in mice and lemmings, why do fish commonly travel in large, relatively dense schools? (2) Why do birds flock together and why do large mammals such as elk and wolves aggregate into herds and packs, respectively? (3) Why does a species persist if food is strictly limiting for that particular species on a cyclical basis? "Strictly limiting on a cyclical basis" means that a resource is essentially nonexistent for periods of time.

Adaptation to resource shortage is, of course, response to k selection. Considering the Lotka-Volterra equations, it seems apparant that adaptation should take the avenue of increased efficiency of resource usage. One way to become a more efficient user of a food resource is to evolve feeding habits maximizing the amount of energy gained per quantity of available resource. This kind of adaptation is discussed later under "How to be a successful predator," p. 116.

Assuming that no hunt is involved in food procural and that capture of a food item mainly involves picking it up, there are still ways to increase intake of usable energy. For seed eaters such means might involve choosing seeds with seed coats that are easily penetrable.

Specialization on a particular size range of food units is also common. The extent of such specialization depends on the availability of food of the desired size. If seeds of a particular size are uncommon in an area, it is unreasonable to expect that organisms will choose to search for them anyway, to the exclusion of other more common types. A common resource is often the choice of a specialist consumer.

Often specialization is nearly a necessity. For example, in studies of marine zooplankters it was found that *Rhinecalanus nasutus,* a relatively large copepod that consumes planktonic algae, can feed on large diatoms much more efficiently than on smaller ones. When given a choice of large and small diatoms, *Rhinecalanus* will feed preferentially (and more rapidly) on the larger species. In contrast, *Calanus cristatis,* a closely related but smaller copepod, preferentially consumes smaller algae. The selective difference between these two species is, of course, the size of their feeding apparatus. It is easier for the larger species to catch larger diatoms than it is for the smaller one. Also, the larger diatom preferred by *Rhinecalanus nasutus* contains nearly five times as much carbon per organism as does the smaller one. If this ratio of carbon can be considered useful energy, then we can conclude that any herbivore able to consume larger food should do so. In this case such consumption, assuming equal numbers of various kinds of planktonic algae, should be at least five times more efficient than more general feeding. Generally, feeding should be specialized whenever specialization confers increased efficiency and increases the amount of energy assimilated per unit time.

Energy utilization

We can begin a short introduction to pertinent theoretical and physiological concepts of energy utilization by assuming than an organism's life can be divided into a number of physiologically distinct stages during each of which food may be limiting to some extent. Next we note that an organism can assimilate food only to some maximal level per unit time. Finally we assume that any energy assimilated may be used in three ways: for somatic growth, for bodily maintenance, and for reproductive growth (gamete production). The way in which energy is partitioned to these three functions is critical to the fitness of the individual. Remember that fitness is measured in terms of the individual's relative contribution to future generations, or as the number of progeny that live to reproductive age. Now recall what Chapter 3 taught us about fitness: It is measured in real terms as the rate of increase of the individual or population. The components of rate of increase (development time, survivorship, and fecundity), are all dependent on how assimilated energy is partitoned. If energy is directed to

gonad development early in life, survivorship will be poor, and development time will suffer. In general, energy should be devoted to somatic growth and development to the extent that mortality and development time are reduced.

Once reproductive age is neared or reached, an optimal energy budget should shift toward favoring gamete production. For species suffering intense food limitation this shift in allocation of energy should be absolute. For example, many insects feed ravenously and grow rapidly as larvae, then cease growth entirely following the final molt to adult form. At this stage all energy is essentially devoted to reproductive processes. The female mosquito of some species, for example, feeds on blood only once so that she may produce mature, viable eggs. In many stream insects the adult stage exists only for the purpose of disseminating eggs. Many of these organisms do not feed as adults. Energy stored during larval stages serves reproductive functions, including flying, mating, and oviposition.

Many species, notably fishes, have indeterminate growth; growth is continuous throughout life. Sometimes there is a substantial shift in energy allocation early in life when some reproductive function begins. Table 4-2 shows the energy budget of Pacific sardines. Notice that no energy is devoted to reproduction during the first year of life; after reproduction begins, however, increasing proportions of energy are used for reproduction and maintenance. It appears that once reproductive size is reached it becomes favorable for these fish to reduce mortality (greater maintenance) and to increase reproduction at the expense of growth. Mortality rates among young fish are high, but those among adult fishes are relatively small. Some fish live and reproduce for many years. During these years, body size and egg production steadily increase. Such species follow a conservative reproductive strategy for the reasons mentioned in Chapter 3. By contrast, salmon and steelhead trout devote all energy to growth for 3 to 4 years at sea. Following this pelagic life cycle stage, large adult fish migrate into fresh water streams

TABLE 4-2. Energy budget for the Pacific sardine

| FISH AGE (YEARS) | WEIGHT (GM) | PERCENT ASSIMILATED CALORIES | | |
		RESPIRATION	GROWTH	REPRODUCTION
0-1	19	81.5	18.5	0.0
1-2	56	89.5	9.8	0.7
2-3	106	92.5	6.5	1.0
3-4	140	95.8	3.0	1.2
4-5	165	97.0	1.8	1.2
5-6	180	97.9	1.0	1.1

*Data excerpted from Lasker, 1970.

to spawn. During this final stage of life, energy stored during the pelagic growth stage is channeled into gonad development and is used for travel upstream to spawning grounds. Some species of salmon that die at the end of spawning do not feed during the run. Perhaps when spawning runs were tremendous, food was too scarce in streams to make feeding worthwhile.

In the Oyashio water off Japan the copepod *Calanus cristatis* has abundant food during early parts of its life cycle, but then is carried south to the point where the Oyashio water mass is submerged and covered by the warm Kuroshio current, which is poor in algal growth. The copepods are then severely food limited. Omori has found that in apparent response to this periodic nutrient shortage the species has a specialized life cycle. *C. cristatis* has in common with other free living copepods a number of preadult life cycle stages. In some invertebrates the importance and durations of these stages are nearly equal. In *C. cristatis* stage copepodite V is greatly extended. This stage apparently is one of energy storage, because feeding is intense and fat deposition is rapid and extensive. Animals are nearly of adult size during this stage, so that a large proportion of assimilated energy can be stored to be used later in reproduction. This stored energy is used while animals are submerged in deep, algae-poor water. These same adaptive modifications of the life cycle are important in the nutrient-poor times which follow the spring phytoplankton bloom in the Bering Sea since the phytoplankton on which these herbivores feed are present in appreciable numbers only during a few months each spring.

It is thought that when feeding has been poor during the copepodite V stage, these animals mature to become males. Male *C. cristatis* are smaller than females and presumably need to devote less energy to reproductive function. Such energy-dependent sex determination seems adaptively sound since only animals with good nutritional status will serve as females. However, without more knowledge of the reproductive behavior of these organisms it is not possible to analyze the effect of a male-biased sex ratio on population fitness.

More is known about the energy utilization strategy of the chaetognath *Sagitta hispida,* a predator of *Calanus* and other similar herbivorous zooplankters. Like herbivorous copepods, *Sagitta* suffers periodic food shortages, which often occur during reproductive times; also like *Calanus,* at least a part of *Sagitta's* strategic reaction involves differential efficiencies of assimilation of different nutrients. Table 4-3 was compiled from data taken on nutrient assimilation of *Sagitta* fed a standard diet and raised at a variety of temperatures. Efficiency is best at 16° C. Increase in temperature decreased metabolic efficiency. For the two sets of dry weight data efficiency loss is great when only somatic growth is counted; when gonadal contribution is taken into consideration, the efficiency drop is much less. These data indicate a partitioning of

TABLE 4-3. Efficiencies of utilization of nitrogen (N) and carbon (C) by *Sagitta hispida**

TEMPERATURE	EFFICIENCY WITHOUT EGGS†		EFFICIENCY WITH EGGS†	
	DRY WT (C)	N	DRY WT (C)	N
16° C	− 31.0	9.1	8.3	53.8
21° C	− 48.3	− 8.8	− 13.9	37.4
26° C	− 26.5	− 2.8	3.8	31.6

*Data from Reeve.
†In terms of percent utilization. Negative values designate error and exceedingly low values.

consumed energy; materials are used more efficiently in reproductive function than when devoted to somatic growth and maintenance. This is metabolic adaptation.

The extent of such metabolic adaptation becomes more impressive when we compare efficiency of nitrogen use to that of carbon (dry weight). Nitrogen assimilation efficiency increases when gonad material is taken into account. However, when nitrogen and carbon are compared more closely, it is apparent that nitrogen efficiency is not significantly reduced by temperature increase although the efficiency of carbon, even with gonad material considered, is less at high temerapture. The use of nitrogen is more efficient and the contribution to efficient use of nitrogen by gonads is far more pronounced than is in the use of carbon. Nitrogen is of overriding importance to reproduction, since it is a major constituent of DNA and newly synthesized proteins. Thus the metabolic adaptation illustrated by this species is extremely refined. Not only is reproduction more efficient in general, but also use of nutrient of particular importance to reproductive function is more highly evolved than is that of more generally useful materials.

A spatiotemporal feeding strategy

Time of feeding of herbivorous zooplankton is an adaptation to diurnal differences in food concentration. *Calanus cristatis* feeds in surface waters at night. MacAllister found that the animals feed at maximum rates upon first entering surface waters. They then become quickly surfeited and eventually sink back to deeper water where they spend the daylight hours. The species feeds maximally when individual algal cells are preparing to divide after photosynthesis has gone on all day and before energy can be respired by the algae. Thus feeding occurs when the algal cells are at their maximum daily size and energy content. The herbivores feed far more efficiently *just then* than at any other time of day.

Copepods, particularly those living in northern waters and those in far southern seas, practice vertical diurnal migration. Vertically migrating species concentrate within specific bands of light intensity. These species rise as the intensity bands rise through the water column, and they sink again during daylight (Fig. 4-2). Some biologists believe that by moving to warm surface waters at night where food is dense and easily captured, individuals can feed and assimilate energy and material at maximally efficient rates. By moving to cold deep waters during the day these organisms are able to conserve energy in a nutrient-poor environment. Energy is used less at low temperatures, and feeding is more efficient at high temperatures. These observations tend to support an efficiency hypothesis. In opposition to this idea is the suggestion that moving up and down in a water column (up to 120-150 meters) must be energetically expensive. To what extent such movement would refute the energy hypothesis remains to be seen.

Kerfoot used data on diurnal migration of the copepod *Calanus finmarchicus* to construct a model of

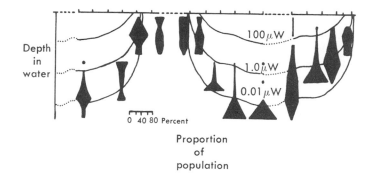

Fig. 4-2. Vertical migration of *Calanus finmarchicus,* a marine copepod. Migration is shown for a large time scale but it also occurs daily, perhaps in response to light quantity and quality. (From Kerfoot, W. B. 1970. Bioenergetics of vertical migration. Amer. Natur. **104**:529-547.)

response to light intensity. He compared this model with the known distribution patterns of vertically migrating copepods. In Fig. 4-3 noontime distributions of *Calanus* are compared with three possible expected distributions: (1) diurnal feeders, which remain in surface waters all the time; (2) mixed strategists, which distribute themselves throughout depth in accord with potential energy distributions; and (3) nocturnal strategists, which follow a band of light intensity that is deep underwater during bright sunlight but at the surface at night. The figure should convince you that this species is a nocturnal strategist. We now need to determine why this is so.

In constructing his model Kerfoot used the facts that light decreases in intensity as it passes downward through a body of water and that the same light intensity that occurs at noon at 130 meters can be found at the ocean's surface on a moonlit night. Reasoning that herbivorous, vertically migrating zooplankton might well follow a band of particular light intensity that is optimal for algae during a 24-hour period, Kerfoot calculated energy gained from following such an isolume. He compared the value with the energy gained by remaining in surface waters without regard to light intensity. For any 24-hour period more energy is to be gained by remaining in surface waters. However, there is an apparent advantage to following

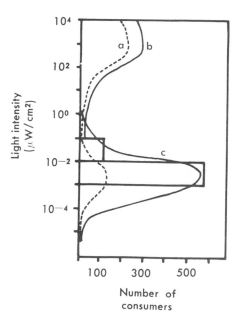

Fig. 4-3. Expected distributions of consumers with depth; predicted on nocturnal, diurnal, and continuous models of feeding. The histogram represents the actual distribution of consumers at midday. This is strong evidence that these consumers follow isolumes of light intensity and productivity in their feeding efforts. (From Kerfoot, W. B. 1970. Bioenergetics of vertical migration. Amer. Natur. **104**:529-547.)

isolumes. Kerfoot considered change in available productivity over the course of a year for various schemes of vertical migration. He found that nocturnal migrators have a nearly stable resource. In contrast, organisms that might orient to water pressure or mixed feeders orienting to light would have a tremendously varied resource. He suggests that resource stability might be of value to nocturnal species. During July surface waters contain a far larger percentage of daily productivity than do deeper waters. One might conclude that living near the surface would be of selective advantage. Kerfoot, however, suggests that this is not so because light intensity is too high near the surface and inhibits many algal species.

Kerfoot's data can be extended to predict the advantages of nocturnal versus mixed feeding depending on numbers of winter and summer months. Kerfoot's investigation shows that July's proportion of total productivity for a mixed strategist (assuming that feeding is at random over depth) is the average of percent productivity with depth of 11%. In contrast, the nocturnal feeder's mean proportion of productivity in July is 0.102. In December mean proportions of productivity for the two groups of strategists are 0.05 and 0.125, respectively, because most mixed feeders are adapted to use high light intensities, but nocturnal feeders can use the low levels of production available in winter months in the North. Assume that July values are roughly representative of summer productivities and that December values represent winter levels. Then assume that polar areas have about 9 months of winter and 3 months of summer per year but that tropic areas have 11 summer months and only 1 winter month. We can then average productivity values for two strategists:

	Mixed	Nocturnal
Polar	0.08	0.1135
Tropic	0.105	0.104

Thus we might expect that mixed feeding will become more advantageous with decrease in latitude. Unfortunately, I know of no data supporting this hypothesis. In addition, the values of productivity used in formulating the hypothesis are open to question, because it is well-known that algae acclimate to changes in light concentration.

Feeding and resource dynamics

The way that members of a population exploit their resources seems to depend both on the population dynamics of the resource itself and on a population's effects on these dynamics. Basically, we can distinguish two kinds of resource: (1) dynamic resources such as insects, whose density is regulated by the feeding efforts of consumers, and (2) static resources such as fruits, seeds, and nuts, which cannot be replaced in response to being consumed. Strategies for use of this latter kind of resource need no special discussion; all conclusions in this chapter apply. However, the dynamic, or compensating, resource is a special case. It is important because it is so common. Dynamics and strategies can be analyzed by a pair of simple equations. Let the resource's growth rate be defined as:

$$\frac{dR}{dt} = R(b - d)\,(1 - R/k)$$

and the consumer's growth rate be defined as:

$$\frac{dC}{dt} = C\{[(R - Rt)\,f] - d\}$$

In the first equation we assume that the resource R has a birth rate b and is limited both by an environmental carrying capacity and by the consumers that impose a death rate, d. The consumer's rate of growth in turn is controlled largely by (1) the amount of resource R available above a lower feeding threshold, Rt, (2) the efficiency with which that resource is harvested, and (3) the consumer's rate of feeding, f. As shown in Fig. 4-4, the consumer can overeat its resource to the point where the resource grows more slowly than the consumer population's needs, or it actually declines because of overconsumption. MacAllister investigated

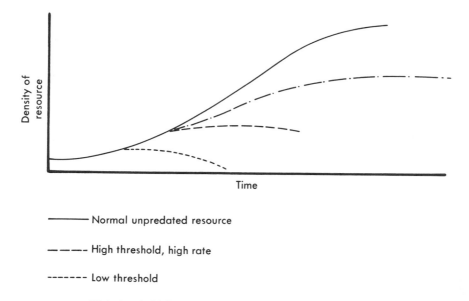

Fig. 4-4. Resource abundance under several strategies of feeding by consumers. The solid curve represents growth of a density-regulated, unharvested resource. "Threshold" refers to level of prey population at which predators begin to feed. "Rate" refers to the rate at which predators capture prey. The predator that begins feeding at a high threshold and does so at a low rate has a replenishing resource. By starting to consume this resource before its capital has built enough to provide much growth a consumer could cause his own demise. This makes little difference in temporary environments such as arctic plankton blooms. Consumers in stable environments should begin feeding at a high threshold of food density and feed at a rate no higher than their food supplies' rate of increase.

the dynamics of a set of equations similar to these. We can derive the following hypotheses from his work. A consumer's feeding strategy should depend on the stability and predictability of it's food source. Herbivores living in decidedly seasonal environments such as the polar seas should feed at high rates and have low feeding thresholds. It makes little difference that they overharvest their resource; the resource does not last long in any event. By contrast, herbivorous zooplankton in stable environments should optimize their feeding by beginning at a relatively high resource density and feeding at slow rates. Their resource is thereby able to maintain high density and, barring unforeseen environmental fluctuation, it should never become catastrophically limiting. Such consumers also benefit from a high feeding efficiency.

Data support the above logic and predictions. Tropical herbivorous plankton have significantly higher minimum food density feeding thresholds and lower rates of consumption than do more opportunistic polar species or races. Development of such strategies may imply group selection.

Resource limitation and sexual dimorphism

Sexual dimorphism in animals serves to increase mating success. It is also an important means of coping with a limiting food resource. If males and females feed in different ways or on different food items, the carrying capacity of the environment can be increased for both sexes, since a wider variety of food is made available. Division can be spatial or temporal, in terms of food particle size and quality. According to data collected by Schoener adult female *Anolis conspersus,* a lizard, eat significantly smaller prey than adult males, and they also occupy smaller and lower limbs of the plants on which they perch and feed. Adult females also feed on smaller prey items than those used by subadult males and juveniles of both sexes. Schoener feels that these differences allow the species to use the habitat more efficiently and to attain population densities greater than would be otherwise possible. Females, whose energetic demands must be greater than those of males and whose survival and function are more important to the population than that of any other group, compete least for resources.

D. H. Morse has shown that there are distinct differences in feeding height between males and females of three species of warbler. Females tend to feed nearer the ground than males in all three species. Could this also mean that females are less exposed to attack by hawks and other predators?

Feeding differences in birds between males and females should be greatest when only females care for young. This follows from the observation that females must spend relatively little time away from their brooding eggs and after their young have hatched, they must make enough feeding flights for their own nutrition and to feed their young. In at least one species of bird sexual dimorphism is of length of feeding flight; females forage nearer the nest than do males.

Sex ratio, mating systems, and *k* selection. The treatment in Chapter 2 of sex ratio suggests that selection against males and reduction of sex ratio is of advantage in *r*-selected species. We need to ask whether *k* selection requires the same or different strategy. The answer to this question depends on energetic conditions. *k* selection seems to imply an advantage to females for producing fewer male offspring. Young females should be better fed and have a greater chance of reaching reproductive maturity than if parents divide their feeding efforts among male and female young.

In many *k*-selected species (particularly birds) monogamy prevails. Reproductive males and females are in equal frequency at mating time. This is particularly true when both parents feed young, which is common in passerine birds. Since females that must feed progeny without the aid of a mate are more stressed than those with mates, an adult sex ratio of $1:1$ should be strongly selected. Therefore, monogamy and equal sex ratio are expected in *k*-selected species when both sexes contribute to the care of young. Most birds and carnivores are monogamous, probably for this reason or because males bring food to their nurturing mates.

However, many mammals are polygynous. Females greatly outnumber males and harem mating systems are common. Males play little role in the care of young. Excess males burden the population because they require resources which females could otherwise glean.

Competition

The study of interspecific competition can be explored on many levels of complexity ranging from simple studies of exclusion in protozoa to far more advanced endeavors that attempt to measure the results of competitive selection on the multidimensional niche spaces of interacting species. In this section we will begin at the simplest level and by reviewing a series of experiments and field observations, we will build our understanding of species interactions to a more detailed and advanced level.

The experiments of G. F. Gause showed that when two similar species are placed in a simple, rigidly defined medium, only one species is likely to persist. His analysis went much farther than this simple observation, however. In a series of experiments various two-species combinations of predacious protozoa were placed in small vials containing nutrient medium and known initial quantities of bacteria on which the predators could feed. Two of these many experiments proved to be of great interest. In the first, *Paramecium aurelia* and *P. caudatum* were grown in the same vial. Sixty-five *P. aurelia* and 25 *P. caudatum* were recovered at a point when the protozoan population seemed to have reached a numerical equilibrium (the carrying capacity of the vial.) Upon growing each species separately, Gause found that the same quantity of medium would support either 105 *P. aurelia* or 64 *P. caudatum*. Considering these results, he reasoned that *P. caudatum* must be energetically or ecologically equivalent to 105/64 *P. aurelia*, and similarly that 1 *P. aurelia* must be equivalent to 64/105 *P. caudatum*, since the medium would support unequal numbers of the two species.

Such reasoning allowed him to interpret mixed culture results which otherwise would have defied analysis. Gause concluded that the 90 total animals in the mixed species culture (consisting of 65 *P. aurelia* and 25 *P. caudatum*) must have been ecologically the same as $65 + 25 (105/64) = 105$ *P. aurelia* units. Thus he was able to show that the species were exactly sharing the medium's carrying capacity. Such results can be represented by a simple extension of the logistic growth equation. Remember that

$$\frac{dN}{dt} = rN(1 - N/k)$$

for a single species using a medium having carrying capacity k. If two species are using the same medium, then the equation logically expands to the pair of equations

$$\frac{dN_1}{dt} = r_1 N_1 \frac{(k - N_1 - N_2)}{k} \quad \text{for species 1}$$

$$\frac{dN_2}{dt} = r_2 N_2 \frac{(k - N_2 - N_1)}{k} \quad \text{for species 2}$$

if the two species may be assumed to be ecologically equivalent.

If, however, as Gause found, one species *(P. aurelia)* uses less medium than the other, it becomes necessary to modify the set of competition equations by adding a term, α, to designate the relative efficiencies or food utilizations of the two competing species. Therefore:

$$\frac{dN_1}{dt} = r_1 N_1 \frac{(k - N_1 - \alpha_{2,1} N_2)}{k}$$

and

$$\frac{dN_2}{dt} = r_2 N_2 \frac{(k - N_2 - \alpha_{1,2} N_1)}{k}$$

The competition coefficients (α) may be determined experimentally from knowledge of the two competitors' rates of increase and the carrying capacity of their environment.

We can interpret the above equations as follows: The realized rate of increase of species 1, dN_1/dt, is equal to its size at any time multiplied by the ratio:

$$r_m \frac{(k - N_1 - \alpha_{2,1} N_2)}{k}$$

and as $N_1 + N_2$ approach k, the realized rate of increase for species 1 declines. If an individual of species 2 uses less medium than one of species 1, the growth rate of species 1 will not be as quickly depressed in a competitive situation as it would be if it were growing in a pure culture started with the same number of individuals. Of more interest, however, is the observation that if the competition coefficient of 1 on 2 is different from that of 2 on 1, the realized rate of increase of the species with the higher, or better, competitive ability will eventually exceed that of the poorer competitor. In the simplest of environments, the poor competitor's rate of increase would eventually become negative and a pure culture of the better competitor will result. This is assuming that the competitors have equal intrinsic rates of increase.

In Gause's experiments *P. aurelia* invariably was competitively superior to *P. caudatum*, but this result could not have been predicted from knowledge of ecological equivalence alone. If we solve the competition equations at equilibrium (when dN_1/dt and dN_2/dt are equal to 0, and $N_1 + \alpha N_2$ and $N_2 + \alpha N_1$ are equal to k) we can construct the set of graphs illustrated in Fig. 4-5. The straight lines of these graphs are called species isoclines. There is one each for species 1 and 2. Each of the species, when started from some initial number below the isocline, will increase until the isocline is reached. At this point, the species with the higher isocline (the one that uses the food more efficiently and therefore has the higher carrying capacity) will increase at the expense of the poorer competitor. Other changes in numbers are denoted by arrows. In case 1 where the species have unequal com-

petitive abilities and resource utilization abilities (hence unequal k's) the species with the higher carrying capacity and intrinsic rate of increase is expected to displace its competitor. In Fig. 4-5, B, the k's differ, but the α's are the same, and there is an equilibrium point at which the species coexist in proportions equal to the ratio of their k values.

This sort of analysis is simplistic and is valid only when species share a common resource completely. In another experiment Gause tested the competitive relations between P. aurelia and P. bursaria. For this

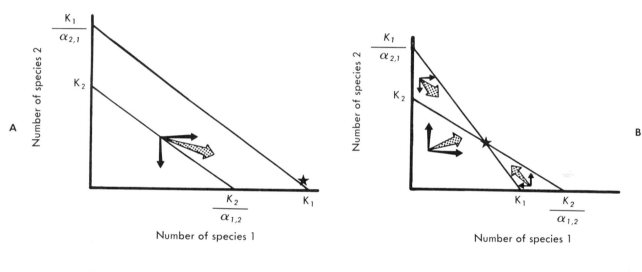

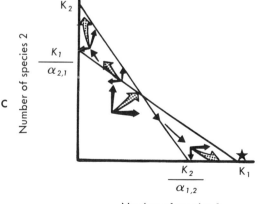

Fig. 4-5. In **A,** species 1 is the stronger competitor and survives to the exclusion of species 2. In **B,** species 1 and 2 both increase when small and their competitive effects balance at an equilibrium point determined by their relative competitive strengths. In **C,** both competitors are capable of excluding the other. Results depend on initial relative densities. Whichever species is more abundant "wins." Arrows denote direction of change of the two species systems (for example, in **A** species 1 increases, species 2 decreases, and the system tends toward all species 1, shaded arrow). Stars denote possible equilibria.

experiment the species were found to coexist indefinitely. Further analysis revealed that the species have different feeding habits; *P. aurelia* feeds in the liquid medium, but *P. bursaria* feeds primarily in the detritus at the bottom of the culture flask. These two sets of results lead to the statement of the competitive exclusion principle: *Two ecologically identical species cannot coexist*. Corollary to this is the observation that species should be able to coexist in nature if, while sharing a resource, they do so incompletely (as in the experiment with *P. aurelia* and *P. bursaria*).

This corollary to the competitive exclusion principle has been the basis of much subsequent analysis of competitive interactions. Park and his students have done much to further the state of competition analysis. Among the many experiments Park performed, the results of two lines of analysis are of primary importance. Park's work has been primarily with members of the genus *Tribolium,* a flour beetle that may be grown in the laboratory under well-defined culture conditions. Beetles are commonly cultured in small vials containing constant quantities of whole wheat flour at constant temperatures and humidities. In one set of experiments Park allowed *T. confusum* and *T. castaneum* to compete at various combinations of temperature and relative humidity. These treatments and their results are outlined in Table 4-4. Results depend on physical culture conditions; at low temperatures and high humidity *T. confusum* is usually the ultimate winner of the interaction. *T. castaneum* wins at high temperature, and neither species is clearly the better competitor at intermediate temperature and humidities. Clearly, *T. castaneum* has the higher rate of increase in hot environments, whereas *T. confusum* is better adapted to low temperature.

This analysis is unnecessarily simplistic. King and Dawson have recently repeated this experiment but with one important modification. As culture containers they used flat pans in which two different gradients of environmental conditions were provided. They built two axes into their culture containers. Along one dimension they allowed temperature to vary; at right angles to this a gradient of food particle size was built. Then experimental beetles were introduced into the center of this container. *T. confusum* moved immediately to the cooler corner in which flour grain size was small. *T. castaneum* migrated to the warm corner of the tray, which contained coarsely ground flour. The two species of flour beetle thus exercised a preference for living conditions. Results like these are a good basis for competition analysis and at the same time allow us to introduce an extremely important concept, that of species niche.

The ecological niche. The species niche has historically been defined in various ways ranging from

TABLE 4-4. Summary of results for competitive interactions between two species of flour beetle*

TREATMENT			
TEMPERATURE	REL. HUMIDITY (%)	WINNER	PERCENT WINS
34° C	70	*Tribolium castaneum*	100
34° C	30	*T. confusum*	90
29° C	70	*T. castaneum*	86
29° C	30	*T. confusum*	87
24° C	70	*T. confusum*	71
24° C	30	*T. confusum*	100

*Data from Park, T. 1954.

"the place where a species lives," to "the species role in the ecological community," to a "multidimensional hypervolume, each dimension of which is an environmental parameter or variable of selective significance to the species." This latter definition, by G. E. Hutchinson, is potentially the most complete and informative. It bears a close resemblance to the extension of Fisher's fundamental theorem discussed in Chapter 2, which suggested that a population should evolve in a way that maintains fitness with respect to all ecological parameters.

Defining the niche as a multiply dimensional fitness space may best be advanced by using a balloon analogy. You already have a mental picture of the fitness distribution of a population as a function of the values of single environmental variables, but refer back to Fig. 2-2. Expanding this to two dimensions we get the sort of distribution shown in Fig. 4-6. We now have a picture of fitness taking into account two differ-

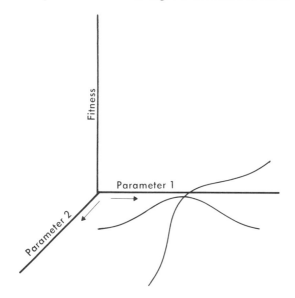

Fig. 4-6. Fitness in two dimensions. Increase in fitness with respect to one parameter may not affect adaptation to another. However, if the parameters are not independent, changes in adaptation of one may be compensated by changes in others.

ent environmental axes, for example, temperature and food particle size. Now think of this figure expanded to many interrelated dimensions as defining the species' fitness in terms of a multitude of environmental variables. It is not unreasonable to visualize a balloon or some similar geometric object. Starting with this model it is only necessary to analyze it in terms of single dimensions. We already know that there will be an optimal value for each environmental variable and the population numbers and fitness should be distributed in approximately normal fashion around this optimal point. We may superimpose many fitness axes. At least some of these will be completely uncorrelated with each other, so that the axes will radiate in every direction from the origin. The result is a sphere, if all the axes are equally long.

The center of the sphere is the point of maximum population fitness, the set of optimal values for all environmental parameters. Just as movement away from the optimum of a one-dimensional fitness distribution is movement toward more extreme parameter values and lower fitness, movement from the center of the multidimensional niche sphere through successive concentric spheres represents movement toward extreme values and lowered fitness. The use of the sphere model contains an implication not possible with the one-dimensional model. Movement toward lower fitness need not be concentric around the origin. Movement toward one pole will change fitness maximally with respect to a few axes, to lesser extents with respect to a few axes, and to still lesser extents with respect to others. Some equatorial axes may be unchanged.

The niche can be thought of as a solid, the center of which is maximally dense (i.e. has highest fitness) and which grades from this area of greatest fitness toward layers of lesser and lesser fitness. The center of this solid is usually the preferred niche of the species, and the entire solid is the fundamental niche. The preferred niche can never by precisely realized in nature, since fitness and degree of preference grade from the optimal region toward the circumference beyond which propagation is no longer possible. Quantification of the species niche can be approached by analyzing properly collected data with multivariate statistical methods. A paleontological study by Green demonstrates methodology. He was able to determine the extent to which several species of mollusc shared niches. (Data collection for such analysis is not straightforward. Colwell and Futuyma discuss possible pitfalls and suggest ways to avoid them.) There are striking niche differences among the age classes of a single species.

The species niche is not equally desirable over its breadth; it may be regarded as a multidimensional fitness distribution with each dimension possessing an optimum as determined by its interaction with other dimensions. Each dimension also possesses values for which fitness is lower. In the absence of competitors, the bulk of any population will tend to occupy the area of niche space that is associated with maximal fitness. This optimal niche area—the species' *preferred niche*—is often contrasted to the *fundamental niche,* which includes all environmental variable combinations in which the members of the species are able to survive and reproduce. (Notice that I just referred to *members* of the species in a way that suggests genetic variability within a population and species as a parameter of the niche. This is perfectly reasonable, since the niche variables are nothing more or less than fitness components.) Encroachment by one or more competitive species on the food or space axes of another should cause all species to change their niche relationships, presumably from preferred, precompetition relations to some other part of the niche space that confers maximal fitness while allowing for the competitive interaction.

Park and King's results can be understood very well in light of this representation of the niche. These results are reproduced schematically in Fig. 4-7. King's results defined what might be regarded as the fundamental niche of the two species. Notice that some overlap exists. Both species will persist in such a system since both are allowed to occupy their preferred, or optimal, as well as their fundamental niches.

 Allowed or operational environment when *confusum* wins

 Fundamental niche of *confusum*

 Allowed or operational niches in Park's indeterminate case

 Allowed or operational environment when *castaneum* wins

Fundamental niche of *castaneum*

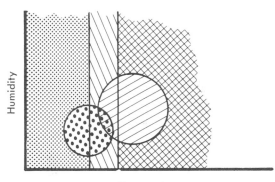

Temperature

Fig. 4-7. Two-dimensional representation of niches of *T. confusum* and *T. castaneum*. Think of these figures as temperature-humidity planes sliced from the unit niche sphere.

Park's experiments restricted the available niche spaces of both species. These niche spaces are represented by shaded areas in the figure. When Park allowed only high temperatures, more of *T. castaneum's* preferred niche fell within the range of allowed environments than did that of *T. confusum*. Therefore, *T. castaneum* had higher population fitness and was able to out-compete the confused beetle. At low temperature the reverse was true. The intermediate temperature restricted the fitness of both species equally. Why were the intermediate results indeterminate? Remember that any species is genetically variable. Park's experiments were initiated with small samples of the total genetic variation of the species. In the various tests the genetic compositions and, therefore, the preferred conditions characteristic of the sample populations were different. Although the initial population of *T. confusum* might have been better adapted to culture conditions than *T. castaneum* in one flask, the reverse might have been true in another test.

Since both inter- and intraspecific competition reduce the fitness of species in zones of resource overlap, they are expected to act to change the means and variances of overlapping parameter axes. If competitive selection acts directly on only a few axes of the multidimensional niche, it will necessarily change the overall shape of the niche. (Squeeze a balloon and see what happens.) It retains its volume but its shape is changed. This analogy corresponds to probable niche evolution. If selection for, say, food particle size causes a reduction in food particle size niche breadth, it may also cause a reduction in total food available. Thus *k* selection on one axis of the niche may reduce fitness with respect to that axis and, initially, with respect to the total niche. It is expected that this reduction will be compensated for by expansion of the species niche onto some other axis. (A species whose food particle size range is reduced by competition might expand its feeding times, or range of habitats covered in search of food, or perhaps variety of food species.)

Competitive selection. Several laboratory studies have investigated the results of selection arising from competitive interaction. Fruit flies are commonly grown in small vials containing an inch or more of a specially prepared fly food into which baker's yeast is seeded. Adult flies subsist on live yeast and lay their eggs on the medium's surface. The larvae hatch and burrow into the culture medium where they mature to the stage of pupation (this process takes about 6 days for *D. melanogaster*). Pupae normally are found either on the medium's surface or attached to the culture tube walls. *D. melanogaster* and *D. simulans* have similar feeding habits. In fact, initially they may be identical in any particular culture. However, in competition their feeding habits and resource usage patterns change after a few generations of competitive contact. *D. simulans*

females begin to deposit their eggs at the edge of the food cups and their larvae move to the bottom of the food cups to feed. *D. melanogaster* feeds in the top center of the food. Larvae of *D. simulans* pupate high on the walls of the culture tubes, but those of *D. melanogaster* continue to use the medium's surface. Evolution of differences such as these is referred to as the process of "character displacement." Both species evolve in such a way as to reduce competitive interaction and the loss in fitness that results. Notice that competitive selection has acted on many life cycle stages, perhaps separately. Adult flies no longer lay their eggs at random on the medium's surface, larvae employ different feeding behaviors, and pupation sites change.

Since competition results in fitness loss to both species we might also expect modification in the life tables of competing species. Nowhere do we find the complexity of population genetics and population dynamics illustrated more strikingly than in the following study. Barker and others studied selection in the competitive interaction between *D. melanogaster* and *D. simulans* and analyzed demographic modifications. He might have expected *D. simulans* to speed up its development time, thereby preempting the resource. However, shortened development time was countered by lower survival and fecundity. Other measures and indices of fitness also showed poor correlation.

These results suggest a general principle. Competitive selection must often be organized on an age-class by age-class basis for both intra- and interspecific interaction. Since selection may be completely uncorrelated among age classes, we expect each to evolve in its own way. Age-class–specific evolution can be in many directions, all of which obey the original competition equation:

$$\frac{dN_1}{dt} = (1 - \frac{N_1 - \alpha_{2,1} N_2}{k}) \, r_m N_1$$

Most obviously a species might evolve a more efficient way of procuring food or other resources, but this is just a stopgap. True avoidance of competitive selection must involve separation in niche space.

The evolutionary value of selection is always maximized at the age of first reproduction. After that, selection becomes less effective in changing gene frequencies, because selected individuals have already had some chance to reproduce. Competitors ought to be ecologically most different at or just before age of first reproduction. By the same logic and considering that food generally becomes less dense as larger particles are required, we might conclude that intraspecific competition can be avoided by dividing into age classes, each of which consumes different kinds and sizes of food items. This is a common phenomenon but we have no way of knowing whether or not the segregation into age classes arose because of intraspecific selection pressure.

Carr of the University of Florida has recently completed a study of the feeding relationships of juvenile fishes. For all of the species common to marine estuaries of Florida's Gulf Coast, foods change with fish size and age. Both species in Fig. 4-8 begin life as plantivores, then one progressively specializes on polychaete worms while the other becomes a generalist, then a vegetarian. For many fish species, food particle size increases with growth. Such changing of food habits and the implied changes in ecological niche may be important to these fish for two reasons: First, by feeding on progressively larger particles of food the growing fishes tend to maximize their efficiency as predators; second (and more obvious), subdivision of the life cycle into numerous age classes, all of which differ in terms of feeding ecology, serves the important function of reducing intraspecific competition among age classes, thereby raising fitness.

Obviously, if all age classes of a species were to consume the same kind of food and if that food source were limiting in quantity, the population size of that species would be limited to the carrying capacity of that one kind of food. In addition, resources (the other kinds of food present in the environment) might well be wasted. Buy why shouldn't members of each age class simply consume a variety of food species or types? The answer to this question will be considered in more detail later, but for now let us make the suggestion that such feeding catholicity would reduce feeding efficiency and, therefore, would lower the environment's effective carrying capacity. This is the ecologist's version of the "Jack of all trades, master of none" principle. A species in which different age classes consume different kinds of food is able to maximize feeding efficiency and amount of available resource per age class, and to maximize ultimate population size, rate of increase, or fitness.

Although competitive selection and character displacement have been studied rather extensively in the laboratory, the phenomenon was first named by Brown and Wilson as a result of field analysis. It is very obvious in some competitive species complexes, particularly birds. A graphic example comes from work by Diamond. In Fig. 4-9 the relative abundance of two bird species from the mountains of New Guinea is plotted against altitude. There is a sharp break in the species composition at 5,500 feet. We assume that these species use roughly similar resources and that the resources grade in composition with altitude, or that some other factor that changes with altitude, such as vegetation composition, temperature, or humidity, is influencing the fitnesses of the two species. The break in relative abundance of the two species is striking. Both species reach their maximal densities at the break point and would be expected to overlap each other's range, thereby competing for resources. Presumably they do not overlap because of interspecific territorial behavior; the species are aggressively excluding one another rather than competing strictly for some resource. Remember that interspecific competition might reduce the fitnesses of the competing species in a region of resource overlap and that anything that will raise fitnesses should be favored by natural selection.

Diamond's work may illustrate two species acting to avoid competitive loss of fitness, but such extreme divergence is not always possible. Species often exist

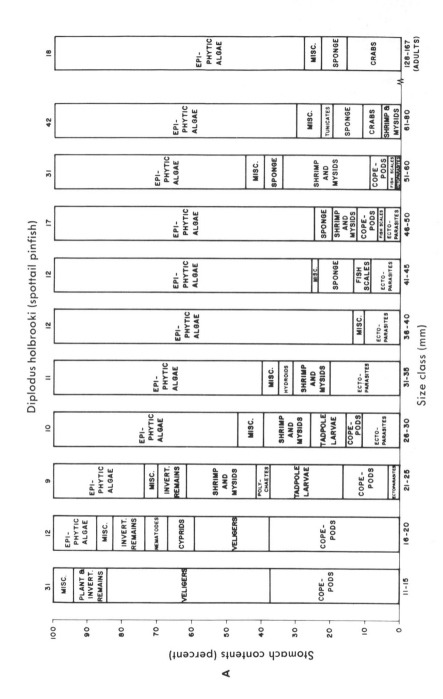

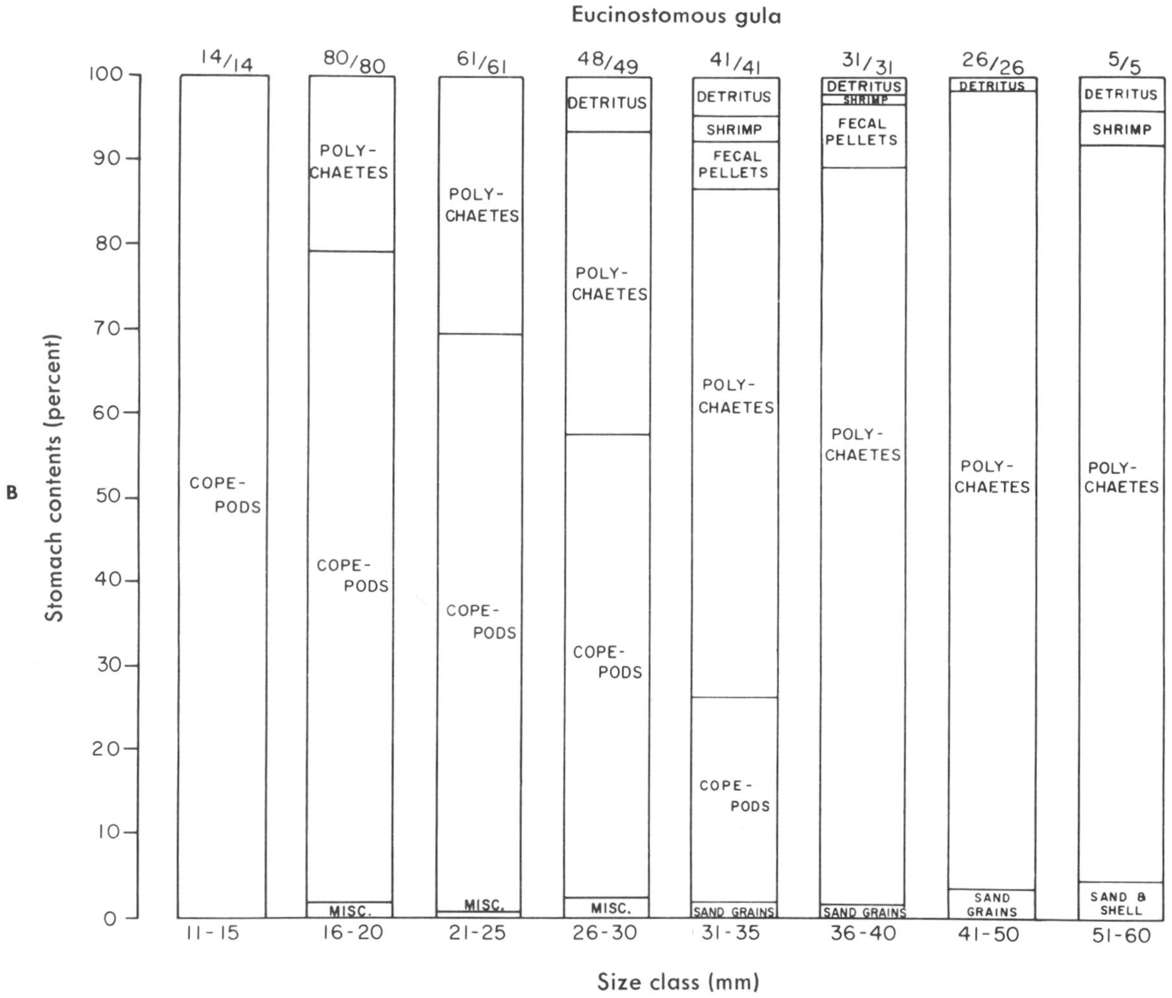

Fig. 4-8. Feeding habits of juveniles of two fish species. **A,** *Diplodus holbrooki* begins life as a feeding generalist, becomes specialized on epiphytic algae, then, perhaps as it outgrows this small plant resource, begins to generalize again. **B,** *Eucinostomous gula* begins as a specialist on copepods and then switches to bottom feeding on polychaete worms. These are but two of twenty species that inhabit estuaries as juveniles. Their changed feeding habits may be the result of interspecific competition for a limiting set of resources.

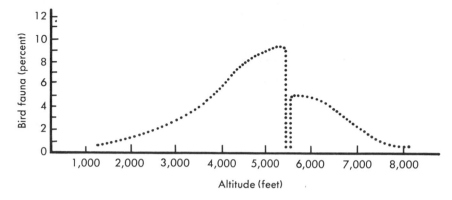

Fig. 4-9. Altitudinal distribution of two warbler species on Mt. Karimui. (After Diamond.)

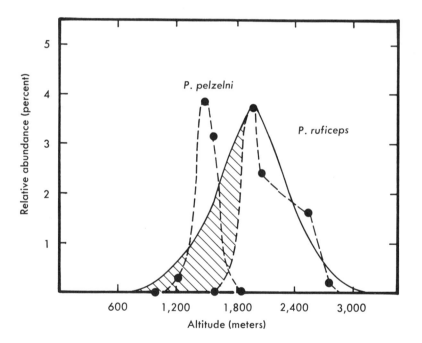

Fig. 4-10. Altitudinal distributions of two flycatchers in Peru. Both species exhibit competitive effects on altitudinal range. Abrupt truncation of the higher species' curve at low altitude suggests competitive exclusion. The shaded area shows extent of competitive expulsion of *P. ruficeps*. The solid curve may represent extent of competition between the two species. The second species is specialized at lower altitudes, may be a more efficient resource user, and should therefore out-compete the more niche-generalized, higher species. High species generality may result from resource shortage. (After Terborgh.)

in dynamic competitive interaction. No field study shows this better than one reported by Terborgh. Fig. 4-10 demonstrates two species of flycatcher which display different reactions to competitive interaction. One of these, *Pseudotriccus pelzelni* (species A), appears to be an ecological specialist; the other, *P. ruficeps* (species B), is a generalist. The specialist's altitudinal distribution is normal, just as we expect of a quantitative trait. In contrast, the distribution of species B is displaced and skewed when compared with a hypothetical distribution which assumes that species B's choice of altitudes is genetically quantitative. If species B had shown a normal distribution, we could then ask: Why do the species fail to diverge in response to apparently heavy selective pressure? The species probably fail to diverge more completely because one or both cannot, due to environmental limitation. Species B occurs at high altitudes and is probably distributed as high as the environment will allow. Species A may also be "trapped" at its present set of elevations because of an altitudinal cline in food character, moisture, or some other environmental variable.

Why, then, is displacement of species B apparently more extreme than that of species A? If altitudinal preference is normal, nearly half of species B's fundamental niche is usurped by species A. This probably happens because species A is a specialist and perhaps is an efficient utilizer of its environment, but species B is more generalized and (theoretically) less efficient. Species A is therefore a stronger competitor than B and B can be displaced.

Why has B not become more specialized, thereby establishing a more defensible niche for itself? This may be the result of resource shortage, the necessity to generalize in compensation, and dispersing effects of intraspecific competition.

Competition-inspired subdivision of the environment is very often not in terms of one primary axis. Since the food-space axes of the niche have many contributing variables (light intensity, temperature, place and time of predator activity, etc.), we might expect that when more than two species are ecologically similar and share an area to such an extent that competition reduces fitnesses, they might diverge from each other on many axes of the niche.

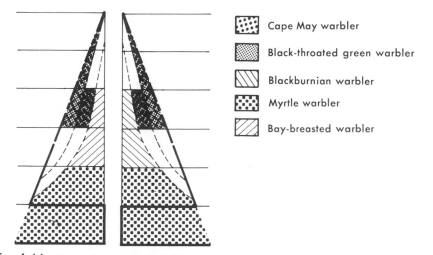

Cape May warbler

Black-throated green warbler

Blackburnian warbler

Myrtle warbler

Bay-breasted warbler

Fig. 4-11. The feeding distribution of five species of warbler in mature white spruce tree. For each species designated zones represent parts of tree in which at least 50% of the feeding activity occurs. (After MacArthur.)

MacArthur has shown ecological divergence or avoidance of competition among five warblers characteristic of white spruce. Some results of this study are depicted in Fig. 4-11. These birds are closely related and their similar morphological characteristics suggest that they might compete heavily. Each species feeds within a different zone of the tree, but zones of the species overlap to varying degrees. This indicates that although zonal segregation of feeding habits has reduced interspecific competition, apparently it has not obviated it completely. However, more detailed analysis indicates that the species have different feeding habits within vegetative zones. Overlapping species use different feeding movements; for example, the black-throated green warbler feeds tangentially, moving across the outer branches, whereas the Cape May warblers from the same feeding zone feed vertically. This is a rather neat way to avoid competition. Tangential feeding probably exposes insects living on the sides of branches, radial feeding exposes those living in limb crotches and on tops of limbs, and vertical feeding exposes those insects living on the undersides of limbs. This behavioral character displacement is nearly complete; the species have diverged and specialized to occupy discrete feeding niches. The five species also nest at slightly different times of year and in different parts of the trees.

Such divergence on more than one subcomponent of the food axis is also shown in Ashmole's study of oceanic terns. Five species of these birds feed on fish and invertebrate larvae at sea and at first might be thought to have similar food niches. However, as Table 4-5 shows, they are divided hierarchically. Although black and blue-grey noddies both consume small fish, they feed at different times of day and so consume different groups of fish, since fish species also have daily periodicity. The brown noddy and the black and blue-grey noddies are separated in terms of a space axis; the former feeds far at sea, but the others stay closer to shore. The black and blue-grey noddies are probably separated by body size. Within this five species complex, three niche axes serve as bases for ecological separation.

Habitat expansion and contraction. Like other population processes, the dynamics and results of competition depend on the individual's environment. Dependence can be so strong that we can speak of physically controlled communities in which little competition occurs. Contrasted with this is the biologically accommodated community, which is physically stable and dominated by interspecific interactions of various kinds. Since competition is significant only when populations are near carrying capacity, those species that are adversely affected by weather conditions must sometimes be free of competitive selection. Also, habitats expand and contract seasonally. Sometimes

TABLE 4-5. Niche subdivision by five sympatric tropical terns

SPECIES	WEIGHT (GM)	CULMEN LENGTH (MM)	BILL CROSS SECTION (MM)	PREY	FEEDING HABITS
Sooty tern	173	41.8		2.4 cm fish	At sea
Brown noddy	173	41.8	21.2	2.4 cm fish	Further out to sea than others
Fairy tern	101	38.7	15.1	< 2 cm fish, squid	At dawn
Black noddy	90.9	43	14.6	< 2 cm fish, squid	During day within 5 miles of land
Blue-grey noddy	45.4	24.8	6.6	2 cm thin fish, minute squid, and crustacea	During day within 5 miles of land

habitat expansion can provide temporary relief from competitive selection.

Zaret and Rand compared the feeding habitats and diets of several species of tropical stream fishes during the dry and wet seasons. During the dry season (when streams consist only of isolated pools) amount of habitat and carrying capacity were both sharply reduced from wet season values. Thus we expect competition to be more intense in the dry season. These investigators showed that in apparent compensation during the dry season the fishes sharply divided the available habitat and food. In the dry season only five species pairs overlapped significantly (60%) in feeding. In contrast, there were eight overlaps during the wet season. The authors note that dry season overlap of diet is probably not biologically significant, because all overlapping pairs fed in separate habitats. On the other hand, some of the wet season dietary overlaps occurred between species occupying the same subhabitats. In cases where overlap was noted in both seasons, it was greater during the wet season.

This work provides excellent (not *a posteriori*) evidence for the selection effect of competition. We do not need to attempt to relate segregation to some evolutionary event long past. Instead, we see niche segregation occurring. The fishes seem to maximize energy intake. All species change their diets between dry and wet seasons. In particular, some specialize when food becomes limiting. By decreasing food overlap with other species, each species probably maximizes net energetic return.

By altering feeding habits these species seem to be maximizing long-term fitness. Retention of low overlap feeding habits would needlessly limit feeding when prey is abundant. Retention of generalized feeding would result in intense competitive loss of fitness when prey is scarce.

To summarize, the competitive exclusion principle proposed by Gause was an important statement historically, but it has needlessly impeded many ecologists who, 40 years later, are still attempting to disprove it. This principle is simply a description of r and k selection and of the relative fitnesses of competing species. In its simplest form it is probably reliable only under laboratory conditions where there is only one resource type available to the competitors. Under natural conditions, competitive exclusion is best replaced by a statement of competitive selection; i.e., when two or more species are in a competitive situation their niches will tend to separate in such a way that the fitness of both is maximized. The form of this separation should be consonant with the niche axis on which the most environmental slack exists.

PREDATION

Although competition is an important phenomenon to the biology of natural populations (some ecologists consider it the most important), predator-prey interactions are also extremely important, particularly in terms of community structure and organization.

Although predators cull surplus individuals from otherwise density-regulated populations, their role and importance are far more general than this. Many mathematical models have been proposed to explain systems of predator-prey interaction. One of these is used later in this chapter. To illustrate, we can now write the classical expression of predator-prey interaction, the two-equation system of Lotka and Volterra. To begin, assume that the prey population grows without limit. That is:

$$\frac{dN}{dt} = rN$$

where r is the prey's rate of increase. To this add a term denoting effect of predation:

$$\frac{dN}{dt} = (r - cp)N$$

where p is the number of predators in the system and c defines the rate at which each predator consumes prey. This is a constant, independent of prey or predator population density.

For the predator population we assume death if there is no prey.

$$\frac{dP}{dt} = -rP$$

states that the prey population will decrease exponentially at rate r unless fed. Introducing prey organisms to the predator, we get:

$$\frac{dP}{dt} = (-r + KN)P$$

where K stands for the ability of the predators to catch a prey and N is the number of prey. This model has a periodic solution. Left alone, the predator-prey system is expected to cycle in relative density of predators and prey.

This model has been severely criticized because of its lack of realism. For example the prey population is assumed to live in an environment that is unlimiting except for the action of predators. In addition, predation efficiencies and prey escape are assumed to be independent of predator and prey densities. Despite the predictions of these equations, typical predator-prey systems in simple laboratory situations follow the course depicted in Fig. 4-12—extinction. In natural systems, this rarely occurs.

According to Rosenzweig, stability of simple predator-prey systems can evolve. In most cases of predator-prey interaction we expect the predator to evolve a prey-catching ability great enough so that the prey population is eventually destroyed. However, we know that in stable communities there must be "prudent predators," which do not eat themselves into extinction. Rosenzweig believes that these predators evolve because (1) the rate of increase of predation efficiency must decrease as predator efficiency increases and (2) because the prey must evolve simultaneously at an increasing rate to avoid exploitation. He shows by analysis of simple differential equations of predator-prey interaction that evolution can result in a stable cycle of correlated oscillations of predator and prey densities. Once this stable cycle is attained by a system, future evolution of both predator and prey results in mutual improvement.

One or both of two conditions are necessary for the persistence of a predator-prey system: (1) some type of "refuge" in time or space in which a segment of the prey population can escape predation, and (2) alternate prey to which the predator can turn for sustenance once his primary prey have reached low densities or retreated to refuge. "Refuge" has many meanings. Refuge can be a simple physical entity, such as burrow or heavy ground cover. Early experiments with two species of Paramecia showed that addition of straw or fine glass tubes to the culture flasks resulted in persistence of both predator and prey species. In this case the individuals of the prey species were small enough to hide inside the glass tubes and the predators were too large to follow.

Huffaker has described a quite different aspect of refuge. He has worked with the six-spotted mite (a pest of orange groves) and its predator, *Typhlodromus occidentalis*. His experiments consisted of exposing a constant area of orange surface to mite attack, presenting this area in patches of different size and number and separated to greater or lesser extent. He found that the system rapidly became extinct when the feeding area was divided into a few large subareas; when the area was divided into a large number of small parcels, the two species cycled in density. These results are shown in Fig. 4-13. The system cycles because the mites were able to seek refuge in both space and time from their predators; they had a much greater dispersal ability and were able to form new colonies as fast as the *Typhlodromas* decimated the old ones. In addition, the herbivores have higher rates of increase than their predators, and therefore, they escape in time (the pests regenerated their populations faster than the predators could destroy them).

Predator choice—search image

Seeking refuge takes one further form when a predator has alternate prey species. Predators commonly switch to secondary prey when the primary, or preferred, prey species reaches low density. In many instances this simply means that the predator uses a series of prey species in direct relation to their abundance, constantly switching to more abundant prey and

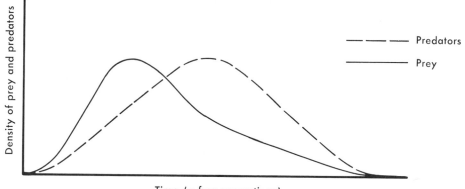

Fig. 4-12. Representation of predator-prey interaction system when no "refuges" are available for either participant.

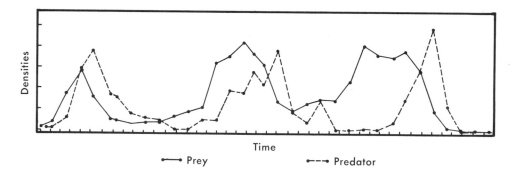

Fig. 4-13. Dynamics of a predator prey system in which the prey are able to refuge from predators. This system becomes extinct after three cycles. (From Huffaker, C. B. 1958. Experimental studies on predation: Dispersion factors and predator-prey oscillation. Hilgardia **27**: 343-383.)

allowing previously preferred species to rebuild their population sizes. Predators form search images for common prey. When such selection by a predator occurs on a single species of prey in which morphological differences exist, it is termed *apostatic* selection. In some snails this is apparently an important force for maintaining morphological diversity within a species. Search image formation is of value to the predator because, by specializing its search for a kind of prey item, the predator soon becomes able to detect it more easily and thereby increases its search efficiency. This is a behavioral attribute which some prey species use to good advantage and which, at the same time is a manifestation of the predator's existence strategy.

Prey refuge—mimicry

Mimicry in prey is the evolutionary response to the formation of search images by predators. It can also be a response to the predator's inability to distinguish prey from other similar (but less edible or completely inedible) objects. Mimicry describes the fact that some prey species have evolved modifications of behavior and form that reduce their chance of death by predation. Mimicry takes many forms, but most commonly the prey species tends to resemble either an inanimate object such as a stick or leaf, or an unpalatable or fierce animal. Some mimetic forms and their models are shown in Fig. 4-14. Some of the prey resemble potentially dangerous animals; others are protectively colored.

In the case of monarch and viceroy butterflies, the monarch is the distasteful model and the viceroy is the palatable form. An optimal ratio is beneficial to both the monarch and its mimic. At the optimal ratio predatory birds sample only a few butterflies before they encounter a distasteful monarch. Once a bird has ex-

Fig. 4-14. Several examples of protective coloration or mimicry. **A,** Mimic model complex of viceroy and monarch butterflies; **B,** a walking stick insect (which must bemuse potential predators); **C,** an innocuous herbivorous insect only about two inches long; **D,** a spider that mimics ants. Two reasons have been proposed for this case of mimicry. The spider may mimic toxic ants or ones with painful bites thereby taking advantage of its model's obnoxiousness. Alternatively, the spiders may be taking advantage of ant density and commonness by hiding in a sea of ants. (See discussion of bait fish schooling.) (**A** from Dillion, L. S. 1973. Evolution: concepts and consequences. St. Louis, The C. V. Mosby Co.; **B** and **C,** photos by Jon Reiskind; **D,** photo by J. Jackson.)

C

D

Fig. 4-14, cont'd. For legend see opposite page.

perienced the taste and after-effect of trying to consume a monarch, it is unlikely to bother either species for some time. When the ratio of viceroys to monarchs exceeds the optimum, the number of trials before sampling a monarch increases. It will now take more than one experience to discourage some birds, since the ratio of good to bad experiences is important in establishing the predator's negative search image. The result is that both species suffer. A bird, having eaten six tasty viceroys in a row is unlikely to be discouraged from further experimentation by one distasteful experience. However, at the optimal ratio, perhaps half of a predator's experience are distasteful, leading to avoidance of both species of butterfly. When the mimic to model ratio is too low, no harm occurs to the system, except that density of the mimic might be below carrying capacity. (See, for example, Brower.)

The monarch butterflies play their own strategic game with their predators. Monarchs are unpalatable because they feed on the milkweed, which produces a cardiac glycoside. This poisonous substance does not affect the butterflies, nor is it detoxified by them. Thus the monarch's protection from predation comes from its diet. Not all monarchs would need to consume milkweed. The same sort of learning response which protects viceroys from predatory birds would protect nontoxic monarchs if any existed. Even so, avoidance of distasteful butterflies is apparently not an inherited trait of their predators. It must be learned anew by each generation of naive predators.

Predator cues

The use of toxic or unpleasant chemicals is widespread among possible prey organisms, but it is asking too much of an animal to emit only toxic or warning metabolic by-products. Quail and other game birds can be scented at great distances by predators attempting to seek them out. Gurin and Carr are elucidating an example of what is probably the unwitting release of predator attractants by shrimp and other marine forage species. Wounded shrimp release short-chain polypep-

tides to their environment. These are detectable in minute quantities and at great distances by predatory fishes and by scavenging crabs. Response to such substances is probably an evolved characteristic of these predators.

Predators also use chemical attractants to minimize search and pursuit times. Many marine fishes travel and hunt in schools. Evidence is accumulating that several of these species emit highly soluble substances when in a feeding frenzy. Such emissions, when detected by fish that have not yet begun to feed, may act as a cue to the presence of prey items and improve the feeding efficiency of the school.

Terrestrial, social predators like wolves announce the presence of large prey with vocal signals. Gulls swoop and soar above a school of fish or other food. This behavior attracts other gulls. These and numerous other behavioral and chemical cues are of great advantage to predators both as individuals and as populations. When prey is scarce but aggregated, predators can greatly enhance feeding efficiencies.

Another use of search images

Prey turn the tables on their image-conscious predators in ways other than mimicry. Pronghorn antelope, white-tailed deer, and several African plains herbivores that travel in species herds have conspicuous white tails or cheek patches. These are of great value when a solitary predator attacks a herd. When the animals notice their predator, they bound off in all directions. The predator is then confronted with a sea of white flashes and erratically bounding shapes. It may become thoroughly confused and give up the chase. The Australian wallaby shown in Fig. 4-15 can flash its white rump patch when fleeing.

A beneficial effect of predators

The action of predators tends to indirectly benefit a prey species complex. The competition equations state that competitive interaction will occur if the population

Fig. 4-15. The Australian wallaby. (Photo by John H. Kaufman.)

is allowed to reach the carrying capacity of its environment. For herbivores, at least, this ultimate carrying capacity is defined by quantity and quality of the food resource. Populations of large herbivores can and do overgraze their range, and in the process they alter the species composition of the plants they feed on. Such alteration of habitat often results in ruin for the herbivore, since nutrient-poor species are allowed to competitively replace preferred plants.

A good example of such three-level interaction is the complex on Isle Royal National Park composed of moose, wolves, and the plant species on which the moose feed. Early in this century moose were introduced to Isle Royale. Since they were the only large herbivores on the island and no predators then existed, the population grew rapidly. In the early 1940s ecologists and naturalists noted that the moose were overbrowsing the island and causing widespread changes in the character of its vegetation. It was then predicted that unless something were done quickly to

reduce the moose population, mass starvation would result, which is exactly what happened.

Following the drop in moose density, which resulted from the population exceeding the carrying capacity of its environment, there occurred an exceedingly cold winter during which Lake Superior froze long enough for a few timber wolves to migrate to the island from Canada. As the wolf population grew, the moose population was kept in check, and more importantly, the browse of the moose began to regain its original character. Control of the moose population by timber wolves lasted until about 1958. We have since learned that the wolf population which had been vigorously growing since its inception reached a maximum density at about this time.

The critical resource that stopped growth of the wolf population was not the moose; their population density has begun to increase again to the detriment of the habitat. The wolf population has stopped growing because the wolves are density regulated. To understand why, we must first know something about the social behavior of wolves. These large carnivores travel in packs, which in large, unlimited areas are usually family units. There is apparently a maximum tolerable pack size beyond which aggressive interactions among males become intense and reproductive success is affected. Just as in the case of the laboratory mice mentioned earlier, neither food nor denning space seems to be limiting for the wolves. Normally, packs are small because young wolves leave after their first year or so to find mates and found their own packs. The probable reason that this did not occur on Isle Royale is that each pack normally has a territory encompassing about 100 square miles. The need for such large territories is apparently a part of the genetically determined behavior of wolves and probably arises because a foraging area of this size is normally necessary to sustain a pack. Isle Royale is only 112 square miles in area and some of the area is composed of a large inland lake. Thus the island can support only one pack, which apparently has a socially determined maximum size.

We have discussed the effects of predators on prey species both in terms of population density regulation and in terms of evolutionary modifications in the prey which reduce the predator's impact. We might next ask more about what it means to be a predator.

How to be a successful predator: the shopping list

It should be fairly obvious that if a predator is capable of reducing the density of a prey species, sooner or later its supply will become "relatively" insufficient. Since most predators are thought to be food limited, the term "relatively" has been the subject of much discussion and theoretical investigation. Pertinent questions to ask about this concept are: (1) At what prey density does the time and energy invested by the predator to capture energy in the form of prey overbalance the gain? (2) Are predators likely to have evolved strategies of feeding to avoid the energy deficit? (3) What are the feeding strategies of predators

and how do these relate to and change with prey density?

The answer to the first question is difficult to determine by itself and is best investigated in conjunction with feeding strategies. The theory of resource usage and the economics of consumer choice as followed by predators can be made quite simple. We start with an energy equation of the form:

$$E_a = E_{PREY} - E_{SEARCH} - E_{PURSUIT}$$

The left side of the equation represents net energy derived by a predator from the capture and consumption of one individual prey. The right side of the equation contains a minimum of three terms that together account for net intake: E_{PREY} (the gross amount of energy contained in a prey individual), and the costs of *search for* and *pursuit of* single individuals of prey. If a predator can realize a net gain in energy by searching for, pursuing, and consuming a unit of a given prey type, he will do so unless some other kind of prey offers higher return for his efforts. A number of points can be gained from this simple mathematical relationship. Search and pursuit time are important to predators. This immediately explains why large carnivores prey on large mammals and why they almost invariably use their energy in pursuit of the very old, the very young, and the diseased.

It is also evident why many predators form search images, taking only the more common items of prey. Given equal pursuit times, their energy is best spent by specialization on an easily identifiable kind of prey for which the search is relatively simple. This is especially true if the search image is extremely common or if food is abundant. Trout commonly feed on only one form of insect during a "hatch" (when particular species of aquatic insect are emerging as adults). At other times these same fish will feed nonselectively. The fact that large trout become cannibals is also readily explained. Capture of a single fish is far more economical than would be capture of an energetically equivalent mass of insects. But why don't small trout follow the same rule, rather than feeding on insects? The answer to this question rests with food availability and search time. Smaller fish are simply more hard-pressed to find other fish that are small enough to eat. For smaller fish other prey are more abundant.

Should large trout eat only other fish, rejecting other forms of food? The answer depends on how abundant the large items of food are. Clearly, if minnows are extremely uncommon, specialization would lead to food shortage. Large items of prey tend to be less common than small ones because they have lower growth rates and lower rates of increase. Therefore, the search time for a large predator, which is energetically best off capturing a large item of food, is high, as are its energy requirements. As a result, larger predators are more catholic in their dietary requirements than are smaller predatory forms. In Fig. 4-16 the goshawk, which is far larger than either the sharp-shinned hawk or the Cooper's hawk, has a correspondingly more varied diet. Shoener has shown that larger birds, whether herbivore or predator, have larger feeding territories than do smaller birds. Schoener's findings are completely reasonable, since the larger birds need more food. In addition, since food for predators is not usually concentrated, more area must be covered by these species in their search for adequate nutrition.

Feeding strategies

Biological time and motion studies suggest the bases underlying many phenomena we notice in nature. Wolves hunt in family units or packs. These large carnivores feed almost exclusively on large herbivores, such as moose, deer, caribou, and reindeer. The capture of large items of food necessitates pack hunting and high levels of social organization. It is unlikely that a lone wolf could successfully capture such prey, but the pack animals, working cooperatively, enjoy reasonable success. Anecdotal accounts by trappers and hunting guides suggest that members of a pack often take turns chasing a fleet prey, each wolf running until it tires, at which time it is replaced with a fresh pack mate. Through such teamwork, the prey is eventually captured.

It is known that large carnivorous fish such as mackeral, bluefish, and striped bass herd schools of their prey in a tight mass from which the entire predator school can easily feed. The reason the bait fish travel in schools is a bit obscure, but it may be to avoid predation! The fitness of a single bait fish must be greater as a member of a vast school than it would be if the single fish swam alone. While schooling, any particular fish's chance of being eaten per unit time is (1/school size) (number eaten from school/unit time). Swimming alone, a sardine is safe only until it is seen by a hungry predator.

Prey density

Reduction in available food results in reduction in specialization. This is certainly reasonable. As all prey become less common, predators might tend to add prey to their "shopping list" for as long as additions continue to reduce search time without increasing the cost of pursuit.

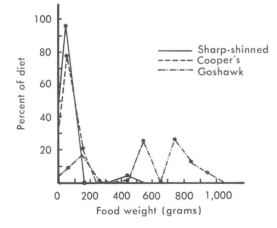

Fig. 4-16. Diets of the sharp-shinned hawk (av wt 135 Gm), Cooper's hawk (av wt 368 Gm), and goshawk (av wt 977.5 Gm). (Data from Storer, 1966.)

Sardines actively choose food type. Very young larvae consume mainly unicellular algae. There follows a life cycle stage during which a choice between algal cells and small zooplankters is possible. It has been shown only recently that this decision is based on the relative frequencies of available food types. When zooplankton exceed 2% of available food items, most juvenile sardines cease feeding on algal cells and begin eating zooplankters instead. The frequency mentioned above is in terms of numbers, but a predator must be strongly affected by food quantity in terms of biomass, or energy. Zooplankters of the type fed the experimental fishes were many times larger than the algal cells involved. Hence the predators must have been cuing on size as well as relative number of food items. That size of individuals is a powerful stimulus is shown in a recent study on bluefish. Investigators found that these fish, having been fed to satiation, would resume feeding if presented with very large items of food. Normal-sized food items were ineffective feeding stimuli.

Baleen whales have adapted in two distinct ways to food shortage. Nemoto, a Japanese biologist, has investigated the head anatomy and feeding habits of these whales and has found that they can be divided into "skimmers" and "gulpers." Skimmers, like the great blue whales swim through the water with their mouths open, capturing everything in their path. Periodically they force water from their mouths, straining out the organisms it contains. These whales feed on zooplankton, which are relatively tiny marine animals (many are less than an inch long). These animals are scattered throughout the water mass. The whales' largest food is a small squid (not the giant squid). Their search efforts are thus generalized to a great extent. The only direction to their quest for food involves feeding near the surface at night when many zooplankters are there.

Gulpers, on the other hand, apparently search out aggregations of prey which they "gulp" up. Their food consists of squid, small schooling fishes such as mackeral, and characteristically aggregated inverte-

brates. These whales search for their food to a greater extent than do skimmers, but they are dependent on finding large concentrations of food organisms.

Of these two forms of whale the skimming baleen whales migrate to warm waters in the winter. The migration is necessary because the prey of this whale is present in polar waters in low quantities during winter months. Tropic waters have fairly large populations of suitable food year around, although zooplankton densities peak in fall and winter. It is likely that whales divide their feeding efforts between two distant grazing areas because were they to remain in one, they would soon exhaust available food supplies.

Perhaps the best example of a predator switching from generalization to specialization was shown by C. S. Holling. He was interested in the dynamics of the predator-prey system, and used for his model system a vole whose foods included three species of insect pupae. Among these insects the pine sawfly is interesting for economic reasons. Using a virus infection, Holling reduced sawfly population density in one plot of forest and then observed events as the prey repopulated the area. During the time of sawfly scarcity, voles fed on other species of insects, earthworms, and any other detectable prey. As the sawfly population grew, Holling noted two responses characteristic of the predators. The first of these he called a *functional* response. Voles established a search image for the increasingly abundant sawfly cocoons and began specializing on them to the exclusion of other prey. The functional response consisted of a sigmoid curve of number of prey eaten per individual predator, as illustrated in Fig. 4-17. The upper limit illustrated for the function response is the point at which the predators are satiated.

Equations have been written to explain Holling's numerical response curve. All of them fail to account for an observed initial lag in the relation between prey eaten per predator and predator density. Tinbergen suggests that the lag in functional response curves of some species may be a threshold learning response; predators may need to encounter a threshold density

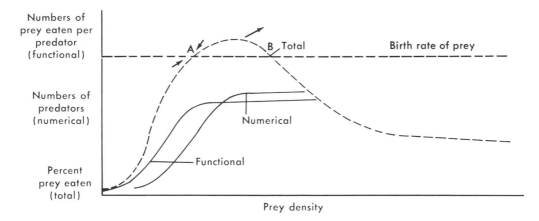

Fig. 4-17. Schematic representations of predator response to increase in density of a prey species.

per unit time of a particular kind of prey before they form a search image and begin specialization. The duration of this lag phase should depend on the relative attractiveness of the previously preferred food and the new, growing food supply. One might expect that a prudent predator would be loath to retool his feeding activities unless his ultimate gain will exceed the effort lost by retooling.

According to Holling, four characteristics of prey affect the functional response of predators. These are (1) caloric value of prey (gross intake of nutrient per prey), (2) prey availability, (3) the prey's stimulus to the predators (search and pursuit time), and (4) the prey's attractiveness (palatability). To test the relationship between strength of stimulus and functional response Holling buried cocoons under different depths of sand in a laboratory experiment. Mice ate fewer deeply buried cocoons than cocoons nearer the surface. (In natural situations cocoons are buried under litter of pine needles and forest duff. This burial has no effect on the predators' detection of the cocoons.)

To test the effect of food quality compared to alternate foods, mice were exposed to different densities of dog biscuits, sunflower seeds, and sawfly cocoons. Sunflower seeds proved superior to dog biscuits. The presence of either alternative food reduced predation

on sawflies, but sunflower seeds were more effective.

Errington defines two types of functional response. In compensatory response predators become selective when prey density exceeds a "threshold of security." Below this threshold individual prey are difficult to find and to remove from secure refuges. Above the threshold, excess numbers of the prey population become wanderers and are vulnerable to predators. In contrast, some large predators like wolves actively and selectively search for prey. Obviously, from what we know about predation and feeding strategies, the extent to which predators cull excess members of a population or search for those at large must be determined by palatability, the extent to which prey stimulate their predators, and the net caloric value of the prey relative to net caloric values of alternate prey species.

Holling distinguishes three basic models of functional response. In the first, increase of predation is linear with increase in prey density. One might expect such a response when prey types are equal in all respects other than density. In a second model number of prey eaten at first lags behind increase in prey density and then grows exponentially until a satiation threshold is reached. Such a response might be expected if the prey are difficult to find and not worth search effort at low density. Errington's compensatory predation applies here. Finally, the response of predators might be exponential with increase in prey density. No lag phase is noticeable. If the chosen prey species is considerably superior to all others and is highly attractive, such a response of predators would be a good strategy.

The second noticeable response according to Holling is the *numerical* response. The voles had been prey limited and rapid population growth followed increased food supply. Vole population growth was eventually limited by the action of some unidentified limiting factor, such as space.

In addition to illustrating how strategies of food utilization change with prey density, Holling's experiment is an example of density-dependent prey regulation. By combining the functional and numerical response curves, Holling derived a total predator response curve that illustrates the dynamics of how predators control density of their prey. This is the composite curve of Fig. 4-17. Percent prey eaten is plotted against prey density; percent prey eaten is the prey's death rate due to predation. Also included in this figure is a line representing the prey population's average birth rate. Further reasoning about predator regulation of prey populations is as follows: Since the total response curve is in terms of prey death rate, any point at which the total response curve crosses the line of average birth rate is an equilibrium point at which rate of increase for the prey population is zero. In the interval between points A and B, d exceeds b for the prey population and, as the arrows show, decrease in density is expected. Whenever the prey population is below density B, control is effective. If, however, the prey's density exceeds the value B, the satiated predators, unable to consume more prey, lose control; percent prey eaten decreases below the birth rate and the prey population expands.

Predation and feeding—an overview

Consumers feed in ways that optimize their net energy intake per unit time or effort. In many cases optimization takes the form of specialization on palat-able, large units or kinds of prey that are easy to find and capture. When density of a resource decreases, two roads are open to the consumer. If some other type of food is available in high density and is otherwise desirable, the consumer will switch to it. However, if no clear choice presents itself, feeding generalization may result. The consumer begins adding items to its diet until further addition reduces benefit to a negative value, as defined by:

$$E_{NET} = E_{GROSS} - (E_{SEARCH} + E_{PURSUIT})$$

Reduction of value can occur because of the decreased intrinsic value of less suitable kinds of prey (perhaps because of small size, or reduced palatability or energy content). It may also be a function of the observation that generalization must reduce searching efficiency when only a few kinds of prey are the object of specialization.

The specialization-generalization strategy can also apply in time and space. As resources decline in abundance (or as the environmental demands on the consumer's energy stores increase) the consumer species may evolve feeding efforts limited to optimal time of day or year, completely giving up the search otherwise. Humming birds, hibernating mammals, and poikilothermic reptiles are cases in point. Some consumers limit their feeding range in times of resource shortage, thereby reducing energy loss due to search efforts. Others, for which search is less expensive relative to the rewards of a single success, increase the size of their search patterns. This is particularly true of carnivores that hunt very large prey. Feeding behaviors are both inherited and environmentally determined.

The evolution of consumers toward higher efficiencies almost implies that evolution of prey will occur as well. Schooling behavior, mimicry, and protective coloration are all examples of prey adaptations that minimize loss to predators. Sometimes it is beneficial for prey to evolve infradispersion. Several tropical tree species whose seeds are consumed by specialist predators spread their seeds over large areas.

This may help to reduce the effectiveness of the predators. Other predator-prey systems have apparently co-evolved into less harmful or even mutually beneficial relationships. The literature of co-evolution and progression from predation to parasitism and eventually to mutualism might bear careful study but is not within the scope of this treatment.

Predator and prey evolutions are cyclic. As predators increase their efficiency, prey increase their abilities to escape. Predator behavior can be used by prey to increase chances of escape, and escape mechanisms can often be worked to the predator's advantage.

PARASITIC INFECTIONS

Just as predators tend to increase in density and percent predation as prey populations become more dense, so also do parasites and infectious disease organisms. Epidemiology is the study of the causes, spread, dynamics, and control of epidemic disease, which is a form of predation important to human populations. We will not have much to say about control of epidemics, but their initiation and spread are of interest because several processes and concepts that we have already discussed are involved.

The course of an epidemic can be described in a series of statements about the probability of an epidemic beginning at various prey (or host) densities. Epidemic diseases depend on host density for their dissemination and spread or, if they require some intermediate host or vector, on the density of the least common of their hosts. The importance of host density is as follows: During our discussion of dispersion we presented the relationship between population density and the chance of a female finding a proper mate. We found that at low densities the probability of success was exceedingly low and that it increased with population density or with clumping. The same statistical reasoning may be applied to the dynamics of epidemics, but we now must consider that there is a threshold density of hosts below which the parasite

population's death rate exceeds its birth rate (or probability of finding and infecting new hosts). Probability of success, or birth rate, of the parasite or disease organism is directly related to the density of its host population. Once the host population density passes this threshold where the infectious organism's birth rate is greater than its death rate (or failure to leave a dying host for a new, healthy one) the epidemic begins.

Black plague, which caused widespread death during the Middle Ages, is infection by the bacterium *Pasteurella pestis*. It depends on there being critical densities of man, rats, and rat fleas. Plague usually struck only in crowded cities. It struck first in the most densely populated slums where rats, people, and fleas were present at above-threshold density.

Plague often claimed the lives of over half the inhabitants of a town. Why not all? To devise an answer to this question, it will be necessary to look at total response. As in the case of predator-prey interactions, total response is measured as percent infection of the host population as a function of host population density. The curve is similar to the predator's total response curve, but its causes and implication are different. The epidemic organism's curve rises as a function of host density and chance that the infection will be passed to a new host, just as does the predator's total response curve. The reason for the curve's inflection point is quite different. In the case of a parasite or disease organism death of a host is tantamount to death of the disease organism. Thus epidemics fail to claim entire host populations for two interrelated reasons. First, as percent of hosts infected increases, total number of host organisms decreases, assuming that infection is synonymous with host death. Fewer hosts mean less chance of infection being transmitted, hence lowered epidemic "birth rate." Second, as percent of potential hosts infected increases, the chance of a disease organism's finding an uninfected host decreases (Table 4-7). Hosts that are already infected are already partially dead. Therefore, by extension a secondarily infecting organism is also nearly dead, since it has infected a dying host which may not offer much opportunity for life cycle completion. We conclude that the death rate of an epidemic organism must increase as the epidemic progresses. Simultaneous increase of death rate and decrease of birth rate eventually cause an epidemic to "run its course."

The above analysis is fine for mortal epidemics, but does it tell us all we want to know about less severe diseases? Organisms infected with a nonfatal infectious disease often develop resistance to the disease organism. Since the proportion of resistant individuals increases with the course of an epidemic, the infectious organism's birth rate must decrease. As before, the epidemic eventually ends. In some cases resistant individuals are carriers of the disease (e.g. typhoid fever) and act to transmit it. In other cases, resistant organisms serve as dead ends for epidemic parasites.

Control of parasitic and bacterial epidemics is often accomplished through lowering population density of vectors or intermediate hosts. This reduces the chance that parasitic organisms will be spread to new hosts, therefore reducing their birth rate and raising their death rate.

In contrast to predators, parasites do not consume their prey. In fact, the best of all possible parasites is one that does its host no harm. Whereas predation often acts to increase reproduction rate in prey species by lowering population density from limiting levels relative to the food resource, a study by Lanciani suggests that parasites can lower a prey population's rate of increase. He analyzed difference in growth rate (development time), mortality, and number of eggs produced per day per female in the marsh treader as related to parasite density per host. Some of his results are shown in Table 4-6. Most striking is the observation that increased numbers of parasites cause drastic increases in development time, which is one of the major components of rate of population increase.

In addition, the parasites appear to decrease the number of eggs laid per host female, to alter the times of the host's reproduction, and to greatly increase host mortality. The overall result of this parasitism is to lower the host's rate of population increse. It is even more interesting that parasite load effects seem to depend on the host's food supply; hosts maintained at low food levels are more strongly affected by parasitism than are those fed more lavishly.

The parasites themselves seem to be limited by host size and food level. Mites taken from low food level hosts are smaller than those isolated from well-fed water striders. It is probable that this study serves as a model for other parasite-host interaction systems.

COMPETITION, PREDATION, AND ENVIRONMENTAL FACTORS

We have now seen that, taken separately, competition (both inter- and intraspecific) and predation are important factors contributing to the regulation of

TABLE 4-6. Effect of ectoparasitic water mites on a marsh treader*

	MITE LOAD DURING PARASITIC PERIOD (AGE 20 TO 30 DAYS)			
	0	2	6	10
Host r	0.099	0.089	0.083	0.059
Probability of surviving through parasitic period	0.96	0.88	0.85	0.54
Age of first reproduction (days)	24.0	26.3	27.3	28.3
Gross reproduction rate, $m(x)$	40.5	38.6	33.2	27.5

*Courtesy Dr. Carmine Lanciani.

natural populations. We will now investigate how all of these factors interact with each other and with physical environmental parameters to influence population size. For many years, marine biologists and biological oceanographers have been interested in the dynamics of offshore communities of phytoplanktonic algae and their immediate predators, the various zooplankton such as copepods and larvae of marine invertebrates.

Of interest in the study of plankton dynamics are the physical variables (intensity of sunlight and solar energy), concentrations of the important nutrients (phosphorus and nitrogen) and their cycles through the environment, population sizes, and productivities of phytoplankton and zooplankton. Population and nutrient interactions are not simply one-way affairs; nutrients (and to some extent energy) cycle throughout the system. Zooplankton not only consume and lower the density of their prey, but also they provide nutrient for the phytoplankton, encouraging further growth which would otherwise be impossible.

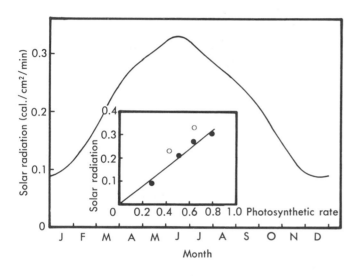

Fig. 4-18. Seasonal variations in average incident solar radiation in the Georges Bank area. Inset shows average observed surface photosynthetic rate plotted against incident radiation. Dots are averages for January, March, April, and May; open circles are June and September. (From Riley, G. A. 1947. Factors controlling phytoplankton populations on Georges Bank. J. Marine Research **6**:54-73.)

Fig. 4-19. Solid line is the seasonal cycle of zooplankton. Measurements of zooplankton volume by the displacement method are treated by a conversion factor (weight in grams = 12.5% × vol. in cc) to derive a rough estimate of the carbon content. Dotted line is the mean surface temperature. (From Riley, G. A. 1947. Factors controlling phytoplankton populations on Georges Bank. J. Marine Research **6**:54-73.)

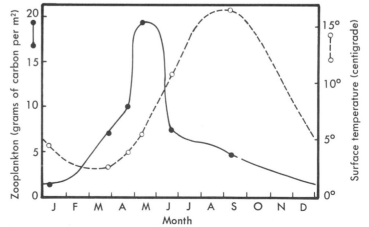

Let us now investigate in turn the factors that are responsible for rapid growth, or "blooming," of marine planktonic algal populations. The first of these is obviously the energy source—incident radiation coming from the sun. As radiation strikes the sea's surface, much is reflected away, but a relatively large proportion penetrates depending on the sun's altitude and incident angle relative to the water surface. Of this, much is either absorbed or reflected during passage from the surface with the result that even in very clear seas, not enough light and energy remain below 300 meters to support growth of plant populations. Already, then, we have one limiting factor defined, and Fig. 4-18 shows the amount of incident radiation in Georges Bank (Atlantic) as it varies with the time of year. Radiation is least during the winter months and increases to a peak in June. A naive guess would be that peak algal population growth would coincide with this peak. That such is not the case is shown in Fig. 4-19.

The peak is in April and is followed by rapid decline until, by June, algal activity is almost at a low point. The reason for this is rather involved. First of all, look at nutrient concentration. The two nutrients that have generally been found to be present in limiting amounts are phosphorus and nitrogen. Both of these are present within the photosynthetic zone of northern seas in more than adequate quantities throughout the winter when light intensities are low and vertical mixing of water is intense enough that algal cells are constantly swept below the "compensation depth" where photosynthesis is no longer sufficient to balance respiration and algal population growth is curtailed because of energy shortage.

As the sun begins to swing north in spring, two things happen. First, incident radiation increases which results in (1) more energy being present at great depth and (2) lowering of the compensation depth at which algal growth can occur. In addition, the in-

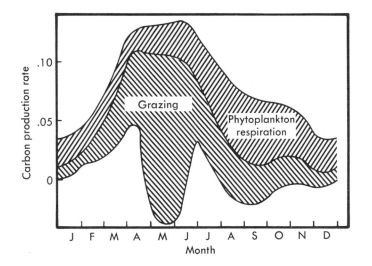

Fig. 4-20. Estimated rates of production and consumption of carbon. Curve at top is the photosynthetic rate. By subtracting the respiratory rate the second curve is obtained which is the phytoplankton production rate. From this is subtracted the zooplankton grazing rate, yielding the curve at the bottom, which is the estimated rate of change of the phytoplankton. (From Riley, G. A. 1947. Factors controlling phytoplankton populations on Georges Bank. J. Marine Research 6:54-73.)

creased radiation warms the surface waters causing a decrease in water density and the formation of a defined layer of equal density water below which vertical mixing of water and algae cannot occur. Thus the algal population is kept above compensation depth starting in mid-April; and growth, which is no longer limited by insufficient energy, can begin. The algal population then rapidly grows and might continue to expand until June except that consumption by herbivores and nutrient shortage now become important. These events occur nearly simultaneously. As nitrogen is used in its oxidized form, NO_3, it is changed to waste products. Some of the nitrogen sinks in the form of these waste products below the compensation depth to be lost for a time in the deep ocean. A rather substantial quantity, however, is reused in the form of ammonium, but even this supply is depleted rather quickly.

Phosphorus is assimilated by the algae, but suffers a different fate from nitrogen. During passage through the zooplankton segment of the community, much is converted back to a form that is immediately utilizable by algae. The bloom is sustained for a short time *as a result* of zooplankter consumption of the phytoplankton. Eventually, however, nutrients reach limiting levels and growth must cease. Simultaneously, the zooplankton population, which during winter was limited by the food supply begins to grow in response to a burgeoning food supply. Eventually as shown in Fig. 4-20, the zooplankters begin to consume algae faster that it can be produced and both they and their food supply decline.

Thus we see that algal population dynamics involve mediation by the physical factors (water temperature and energy supply) and control by nutrient supply, competition, and the action of predators; in addition, there is some feedback from the predators, which tends to sustain growth of the algal population.

By early fall sunlight energy again becomes limiting so that the fall turnover curtails population growth rather than sustains it, even though the turnover restores nutrients to the sea surface, reestablishes vertical mixing and supplies nutrients for the next season. The cycling and interactions indicated by this work are common to all natural systems and are the basis for one aspect of community ecology, which is the subject of the following chapter.

BIBLIOGRAPHY

Ashmole, N. P. 1968. Body size, prey size, and ecological segregation in five sympatric tropical terns (Aves: Laridae). Syst. Zool. **17**:292-305.

Barker J. S. F., and R. M. Podger. 1970. Interspecific competition between *Drosophilia melanogaster* and *Drosophila simulans:* Effects of larval density on viability, developmental period, and adult body weight. Ecology **51**:170-190.

Brower, L. P. 1969. Ecological chemistry. Sci. Amer. **220**:22-29.

Brown, W. L., and E. O. Wilson. 1956. Character displacement. Syst. Zool. **5**:49-64.

Carr, W. E. S., and C. A. Adams. 1973. Food habits in juvenile fishes occupying seagrass beds in the estuarine zone near Crystal River, Florida. Trans. Amer. Fish Soc. **102**:511-540.

Christian, J. J., and D. E. Davis. 1964. Endocrines, behavior, and population. Science **146**:1550-1560.

Colwell, R. K., and D. K. Futuyma. 1971. On the measurement of niche breadth and overlap. Ecology **52**:567-576.

Eisenberg, R. M. 1966. The regulation of density in a natural population of the pond snail *Lymnaea elodes*. Ecology **47**:889-905.

Eisenberg, R. M. 1970. The role of food in the regulation of the pond snail, *Lymnaea elodes*. Ecology **51**:680-685.

Emlen, J. M. 1968. Optimal choice in animals. Amer. Naturalist **102**:385-389.

Emlen, J. M. 1973. Ecology: An evolutionary approach. Addison-Wesley Publishing Co., Reading, Mass.

Errington, P. L. 1946. Predation and vertebrate populations. Quart. Rev. Biol. **21**:144-177, 221, 245.

Gause, G. F. 1934. The struggle for existence. Hafner, New York. (Reprinted 1964.)

Green, R. H. 1971. A multivariate stratistical approach to the Hutchinsonian niche: Bivalve molluscs of central Canada. Ecology **52**:543-557.

Holling, C. S. 1959. The components of predation as revealed by a study of small mammal predation of the European pine sawfly. Canad. Entomol. **91**:293-320.

Huffaker, C. B. 1958. Experimental studies on predation: Dispersion factors and predator-prey oscillations. Hilgardia **27**:343-383.

Hutchinson, G. E. 1957. Concluding remarks. Cold Spring Harbor Symposium Quant. Biol. **22**:415-427.

Kerfoot, W. B. 1970. Bioenergetics of vertical migration. Amer. Naturalist **104**:529-547.

King, C. E., and P. S. Dawson. 1973. Habitat selection by flour beetles in complex environments. Physiol. Zool. **46**:297-309.

Krebs, C. J. 1970. *Microtus* population biology: Behavioral changes associated with the population cycle in *M. ochrogaster* and *M. pennsylvanicus*. Ecology **51**:34-53.

Lanciani, C. In press. Parasite-induced alterations in host reproduction and survival.

Lasker, R. 1970. Utilization of zooplankton energy by the Pacific sardine population in the California current. In J. H. Steel, editor. Marine food chains. University of California Press, Berkeley.

McAllister, C. D. 1970. Zooplankton rations, phytoplankton mortaility, and the estimation of marine production. In J. H. Steel, editor. Marine food chains. University of California Press, Berkeley.

MacArthur, R. H. 1958. Population ecology of some warblers of northeastern coniferous forests. Ecology **39**:599-619.

Morse, D. H. 1971. The foraging of warblers isolated on small islands. Ecology **52**:216-229.

Nemoto, T. 1970. Feeding pattern of baleen whales in the ocean. In J. H. Steel, editor. Marine food chains. University of California Press, Berkeley.

Nicholson, A. J. 1948. Competition for food among *Lucilia cuprina* larvae. Proc. 8th Int. Cong. Entomol., pp. 277-281.

Omori, M. 1970. Variations of length, weight, respiratory rate, and chemical composition of *Calanus cristatus* in relation to its food and feeding. In J. H. Steele, editor. Marine food chains. University of California Press, Berkeley.

Park, T. 1954. Experimental studies of interspecies competition in two species of *Tribolium*. Physiol. Zool. **27**:177-238.

Pollitzer, R. 1960. A review of recent literature on plague. Bull. W.H.O. **23**:313-400.

Reeve, M. R. 1970. The biology of Chaetognatha. I. Quantitative aspects of growth and egg production in *Sagitta hispida*. In J. H. Steele, editor. Marine food chains. University of California Press, Berkeley.

Riley, G. A. 1947. Factors controlling phytoplankton populations on Georges Bank. J. Marine Research **6**:54-73.

Rosenthal, H., and G. Hempel. 1970. Experimental studies in feeding and food requirements of herring larvae *(Clupea harengue* L.*)*. In J. H. Steele, editor. Marine food chains. University of California Press, Berkeley.

Rosenzweig, M. L. 1973. Evolution of the predator isocline. Evolution **27**:84-94.

Schoener, T. W. 1967. The ecological significance of sexual dimorphism in size in the lizard *Anolis conspersus*. Science **155**:474-477.

Storer, R. W. 1966. Sexual dimorphism and food habits in three North American accipiters. Auk **83**:423-436.

Terborgh, J. 1971. Distribution on environmental gradients: theory and a preliminary interpretation of distributional patterns in the avifauna of the Cordillera Vilcabamba, Peru. Ecology **52**:23-40.

Tinbergen, N. 1960. The natural control of insects in pinewood. I: Factors influencing the intensity of predation by song birds. Arch. Neerl. Zool. **12**:265-343.

Watson, A., and D. Jenkins. 1968. Experiments on population control by territorial behavior in Red Grouse. J. Anim. Ecol. **37**:595-614.

Zaret, T. M., and A. S. Rand. 1971. Competition in stream fishes: Support for the competitive exclusion principle. Ecology **52**:336-343.

5 ORGANIZATION INTO COMMUNITIES

COMMUNITIES: THEIR STRUCTURE AND FUNCTION

The study of community ecology is really nothing more than investigation of the interactions among species that live together in a particular area or defined habitat such as a field or a pond or a stream. The most important of these interactions are competition and predation. In a community these occur among many species, not just among a few as discussed in Chapter 4. In this chapter we will discuss community structure and function from two quite different points of view. We will begin by expanding previous discussions of competitive and predative interactions, but will treat them as strands in the web of community structure. Thereafter we will discuss structure from the standpoint of energetic relationships—production, import, and transfer of energy among members of the various feeding levels of the community. Finally, we will look at pathways of nutrient transfer.

To get an idea of community complexity consider the web of feeding relationships presented in Fig. 5-1. Such a food web is typical of northern streams and actually represents a relatively simple assemblage of interacting forms. However, notice that even for this simple community there are many sources of energy and nutrients. Consider first the categories green algae and diatoms. Each of these categories contains large numbers of species of single-celled and chained plants that float freely in streams or live attached to rocks and other stream-bottom debris. Each of these species has

its own nutrient and light requirements for maximal growth and, therefore, each occupies its own niche in the stream habitat. These species as a group are called primary producers because through the process of photosynthesis they are able to produce organic macromolecules in which transduced solar energy is stored in the form of chemical bonds. By capturing energy and using it to build organic macromolecules such as sugars and polypeptides, these organisms produce foodstuffs and store energy for the rest of the community. Detritus is primarily dead plant material that has fallen into the stream previously, but nevertheless it is an important source of nutrients.

Above the producer level the community's complexity with respect to interactions begins to rapidly increase. At the first feeding level above the producers are several species of insect larvae, all of which depend on the producers for their nutrient and energetic needs. These larvae function simultaneously as secondary producers (because by eating plants and converting some of their organic material into new forms, they produce the food of other animals) and as primary consumers, or herbivores. The designation ''primary consumer'' comes from the fact that no energy is produced beyond the plants; it can only be consumed in successive stages by higher feeding levels.

In the next step in the food web are other species of insect larvae, but some of these belong to more than one feeding level. The mayfly, for example, consumes midges as well as some of the primary producers. This insect is both a primary consumer and a carnivore.

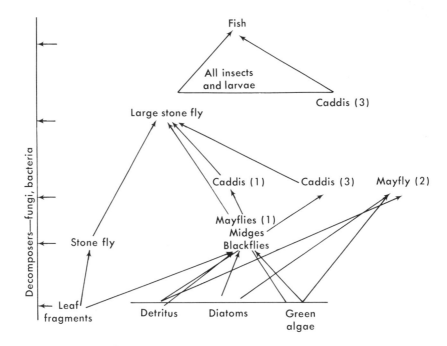

Fig. 5-1. Simple stream community food web.

Some of the insects listed feed only as larvae; therefore, their position, at least in terms of energy transfer, is completely illustrated by this diagram. However, the black flies, which are herbivores as larvae, become carnivores as adults.

Some species serve as food for many other animals. In fact, it is quite difficult to find an animal that has only one role in community function or one place in the community's structure. A further component of this community has not been drawn; young fish may be consumed by some of the larger insect larvae, but as adults the fish consume all of the animals shown in the diagram. Overlying this entire assemblage is a host of decomposer organisms. Among these are fungi and bacteria. Their importance cannot be overemphasized, since their major function is release of nutrient elements such as carbon, phosphorus, nitrogen, and sulfur from the refuse and remains of the rest of the community's inhabitants. In addition, it is becoming increasingly clear that decomposers serve an important role in energy transfer in some communities. More will be said of this later also.

This description may give the impression that community interactions are only in terms of food and energy transfer. Such is not the case, of course. Consider, for example, the flora and fauna of a narrow region of the rocky intertidal zone along the Oregon coast (Fig. 5-2). In areas of this sort several species of two basic kinds of algae are found. Many of these are large, stalked, and cover considerable areas; although they may be eaten by some of the local invertebrate animals, they serve largely as shelter for the rest of the community. Because they often cover the rock face with a resilient, opaque blanket, they probably reduce the force of waves and shelter organisms from direct sunlight. Wave wash might otherwise cause high mortality in animals of the rocky intertidal; similarly, direct sunlight, particularly when combined with wind, might cause death through desiccation.

Sheltered by the fronds of these larger species of

Fig. 5-2. An area of rocky intertidal area on the Oregon coast showing *Pisaster ochraceus, Mytilus californianus,* barnacles, and large attached algae. (Courtesy Peter W. Frank and John Cubit.)

algae are several different species of nemertean worm, fish, several kinds of polychaete worm, three species of crab, limpets, and predaceous snails. An even more diverse assemblage of animal life can be found sheltering among the ''roots,'' or holdfasts, of several of the larger algae. These large algae also serve a second distinctive function. The fronds, continually washed against the rock face by waves, discourage colonization of the rock by many normal inhabitants including barnacles, mussels, and goose-necked barnacles. Such communities are far from homogeneous from place to place; some interactions are restrictive and destructive of other species and serve to establish spatial heterogeneity within an area.

The other important group of plants are the microscopic algae filming the rocks. These serve two purposes. They are food for a variety of snails, primarily the limpets. In addition, part of their activity and metabolism involves production of excretory products that erode the rock face and prepare it for more complex life forms. These algae act as the first stage in primary succession, the process by which communities of organisms continually change their environment and, as a result, their own species composition.

Some animals in this intertidal community, notably several species of barnacles and mussels, derive all of their nutrients from the sea at large. They act as net importers. Their community function is similar to that of the algae, since their existence is important to the rest of the community both from the standpoint of nutrition and because of the shelter they provide for other organisms. They are all filter feeders and consume any material that comes within their sphere of influence. Hence they are a source of mortality for their own larvae as well as for those of other members of the community. The shelter function these species perform is rather important. They occur in large, closely packed beds attached to rocks which have myriad crevices and holes. These encrusting forms of animal life provide retreats where other species are protected from predation and the harsh effects of inclement weather. No fewer than twenty-two species of worms, nemerteans, crabs, molluscs, sponges, and other forms are commonly found sheltered within a single mussel bed. These infaunal species occupy all feeding levels. Some, like the nemertean worms, are predators which prey on polychaetes which, in turn, hunt them! Many of the crab species are plankton eaters or scavengers, but some are carnivorous as well. There are three forms of interactions in any community—feeding, shelter, and competition. As has been mentioned before, many of the species are predatory, and predators are eaten in turn by other larger or more voracious predators. Others are herbivores or detritus feeders.

Of the three barnacle species common in the rocky intertidal zone of the San Juan Islands off the coast of Washington state, *Balanus balanoides* is best adapted to higher regions of the littoral, and it crowds *Chthamalus fragilis* from the rock face wherever the two species occur as young together. The growth rate of *Balanus* is higher than that of *Chthamalus*, which enables *Balanus* to undercut, overgrow, or crush the slower growing form. Relatively little of this competition occurs among individual *Balanus balanoides*. In lower areas of the intertidal, *Balanus cariosus* can outcompete both *Balanus balanoides* and *Chthamalus* in the same manner. This spatial competition also extends to include the mussel *Mytilus californianus,* which attaches itself to the rock in dense clusters. Once settled, *Mytilus* are able to overgrow barnacles resulting in their eventual starvation. Nevertheless, rocky shore is never entirely covered by *Mytilus* for three reasons. First, the mussel is intolerant to desiccation and, therefore, only occurs in lower reaches of the intertidal. Second, wave wash and drift logs continually batter the rocks and tear large patches of *Mytilus* loose, thereby creating open patches that can be recolonized by barnacles or attached algae.

The third factor maintaining community heterogeneity and structure is predation. In Chapter 4 you learned that interspecific competition is meaningful only when animals or plants actually approach the carrying capacity of their environment. For the major intertidal competitors space is the limiting factor, but predators act to reduce population sizes below the critical level.

There are three major predators and a fourth group of animals, limpets, that mediate community function in this assemblage. The limpets are small snails that crawl slowly over the rock surfaces feeding on microscopic algae. They do so by rasping the rock surface with a toothed tongue-like structure, the radula. In the process of feeding, these animals scrape and bulldoze young *Balanus balanoides* and *B. cariosus* from the rock. One or two large limpets can keep a square foot of rock surface quite free of young barnacles. The limpets have a minimal effect on populations of

Chthamalus, because young of this species are much smaller than those of the two *Balanus* species, and they settle in crevices and small holes in the rock. Therefore, limpets can move over most *Chthamalus* without ill effect.

Limpets are the first factor in *Balanus* control. Eventually, however, *Balanus* are left that are large enough to withstand the crowding and bulldozing of the limpets. These larger *Balanus* and the maturing *Chthamalus* then fall within the prey size range of two species of *Thais,* a carnivorous snail that feeds by boring holes in the shells of their prey and then licking the prey's tissues out through the hole. These snails are prey size-specific; as *Balanus* and *Chthamalus* begin to enter their size range, only the larger, faster growing *Balanus* are eaten. These predators also attack young mussels, thereby lessening their competitive impact on community structure.

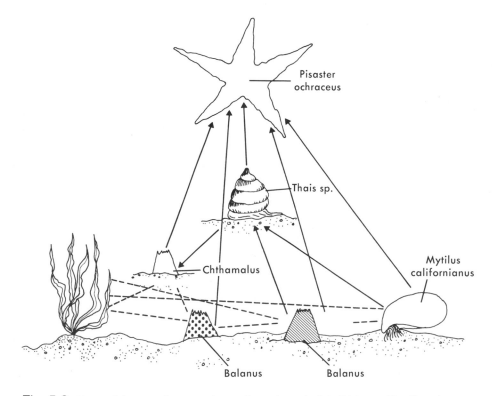

Fig. 5-3. Some of the more important interactions of a rocky intertidal area. The three barnacles, *Mytilus,* and algae compete for space. Competitive interactions are modified by limpets (not shown), which eat young algae and bulldoze young of the three most successful animal competitors from the rock face. *Thais* selectively eats the two most effective barnacle competitors and the mussels. *Pisaster,* a key species, balances the entire system by consuming the most successful of the competitors.

Eventually some barnacles and mussels become too large to fall prey to *Thais*. At this point the predaceous seastar *Pisaster ochraceus* becomes important. These animals are quite selective feeders. Preferred prey are mussels, followed by barnacles, limpets, and other intertidal invertebrates. Because mussels are the preferred prey of this species and because the animals are fairly numerous and have large appetites, they are able to balance competitive interactions among mussels and the competitively less effective barnacles. Fig. 5-3 is a diagram of the above interactions.

The disturbing factors serve in concert to maintain a balance between the space occupied by mussels and by barnacle species. Algae fit into the structure because they can grow in association with barnacles (if they are not too dense). Perhaps the most important lesson of this discussion is that communities are dynamically interacting assemblages of species. Their structure and proper function may well be dependent upon the presence or absence of a single species. In the preceding case the key species was *Pisaster ochraceus*. On Isle Royale the wolf was of key importance. The importance of predators on Arizona's Kaibab plateau was unhappily realized when a massive predator extermination program allowed rapid growth of a deer population, with overbrowzing and widespread habitat destruction. In general, predators seem to be of great importance to community function and stability.

Species diversity

After reading the preceding discussion concerning the structure of two communities, several questions might be asked: "Why are there so many species? Why do many species exist in a community? Why do communities differ from each other in terms of number of species?" Another equally important aspect is, "How are large numbers of species in a community supported?"

Support of species variety. Competitors tend to diverge on a resource so that overlapping of resource usage is reduced; levels of competitve interaction are balanced against the energetic and uncertainty costs of reducing competition by further specialization. Niche axes, particularly those relating directly to sharing of density regulative factors of the environment, are partitioned among the species of any particular feeding level. We will expand this concept to include partitioning of more than one axis at a time, as in the case of the terns investigated by Ashmole (Chapter 4). In communities with more diverse (and therefore more divisable) habitats and with more available axes of habitat there should be more species in the feeding level that directly uses the habitat diversity and, secondarily, in all dependent feeding levels. This concept has been demonstrated in a number of studies.

MacArthur and MacArthur performed one of the first studies specifically designed to test the effect of diversity on fauna. The number and diversity of bird species were compared with the variety of foliage heights available. They found, as shown in Fig. 5-4,

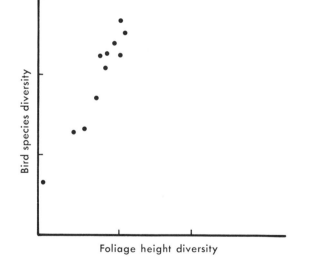

Fig. 5-4. Plot of the relation between bird species diversity and foliage height diversity. (After MacArthur.)

that bird species diversity increases with foliage height diversity, as estimated by the number of foliage heights (i.e. herbs, shrubs, trees) available. Why? First, in any naturally growing forest or vegetated area, the number of plant species generally increases as the number of layers increases, since a species usually has only one growth form in a given habitat and thus can occupy only one layer. These plants provide food of a diverse and subdividable nature both directly (as seeds and leaves) and indirectly (as epiphytic insects). Many insects are quite specific as to vegetation type occupied. Therefore, the more varied the mosaic of vegetation types and species, the more variable will be the insect fauna. Notice that in Fig. 5-4 the increase in bird species is faster than would be expected if foliage height diversity were the only component of the interaction. This can be explained if: (1) foliage type diversity and species diversity increase faster than foliage height diversity, (2) microclimatic stability increases nonlinearly with foliage height diversity, (3) insect and other food type diversities increase with increases in both stability and foliage species diversity.

Diversity builds diversity: Diverse herbivorous insect fauna should foster a diverse class of carnivorous insects. All would be prey for insectivorous birds. A diverse source of food often increases the chances of feeding success of predators. In addition, areas providing more physical diversity are likely to provide more "spaces" in which organisms of various kinds can find refuge as well as food. In fact, both number and diversity of available food types and availability of predators and nesting sites increase as a function of increase in number of foliage layers.

Areas in which vegetation is mature enough to be multilayered will, as a function of this layered structure, be physically more stable in the face of climatic variation. Anyone who has spent some time out-of-doors knows that on a year-round basis a deep woods has far less variable a climate than does an open field. The trees and other tall vegetation act to cut the force of cold winds in winter, and in the summer they provide

shade against heat. The MacArthurs' study dealt with perching birds, but it is likely that there are similar patterns of diversity among other taxonomic groups.

The above facts suggest one reason diverse communities are diverse. They also draw attention to one of the variables that produces differences in diversity among communities—environmental stability. Natural communities vary to a tremendous extent in the number of species that inhabit them. Tropical rain forests are particularly diverse areas; in contrast, diversity is exceedingly low in the Arctic. Natural temperate forests contain many animal species, but agricultural land, such as a corn field, contains only a few species (and these are often serious economic pests!). Fig. 5-5 shows that marine bottom communities vary in a regular way with water depth, increasing in diversity with increased depth. Other data suggest that diversity increases with total energy passage through the community, but this does not hold for all systems.

The problem of why some communities contain few species and others many may well be a complicated one; no simple explanation is likely to define the reason underlying differences in diversity among communities. Several explanations however, have, been proposed. These are all based on observations of geographical variation in community composition. For example, the Arctic land mass has remarkably few species, but the tropics contain a great diversity of plant and animal life. For a time it was thought that this difference was primarily the result of extremely low temperatures of Arctic environments in contrast to the near body temperature of the tropics. The argument proceeded as follows: Most organisms must have originated in near-tropic environments. Even though latitudinal migration of species must have occurred from these origins, few forms of life would be likely to adapt to extreme cold. More careful thought about this point, however, suggests one obvious fallacy of the argument. Given that migrants had some genetic variability for temperature adaptation to start with and given the possibility of recombination and mutation-derived variation, there is no reason to preclude even-

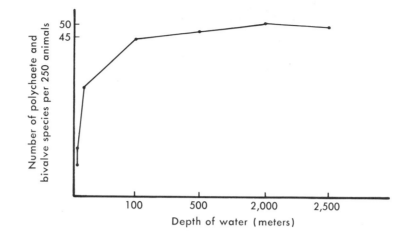

Fig. 5-5. Marine benthic faunal diversity. Diversity at first increases with depth and then decreases. Other information suggests that the fauna become progressively more impoverished as depth exceeds 2,000 meters, perhaps because of decreased food supply at greater depth. (Data from Sanders.)

tual adaptation of species to Arctic conditions. Therefore, although we cannot dismiss the temperature stress argument, we must treat it with some care. Perhaps it is not unreasonable to suppose that its utility and possible application arise from some interaction of average time necessary for successful colonization of a new habitat and the actual amount of time that has been available for Arctic colonization. We suspect, however, that adaptation of species to thermal conditions characteristic of tropic and subtropic areas would have been relatively easy and quick.

If we measure the time since an area was last subjected to a drastic climatic change, it is true that more time has been available for the elaboration of diversity in tropic areas. A grading of the large-scale temperature variation caused by ice ages is easily distinguished from northern to southern latitudes. More stable areas would logically have been characterized historically by less fluctuation-caused species extinction. Hence they should presently contain more species than areas characterized by large-scale environmental instability and variation.

Data on the species diversity and phylogenetic

structure of marine benthic communities support the hypothesis that long-term environmental stability leads to increase in number of resident species. Vinogradova surveyed the geographical distribution of various groups of marine invertebrates. She concentrated her efforts on depths greater than 2,000 meters. Species diversity is greatest at 2,000 meters and declines rapidly with increased depth, probably because of nutrient shortage, which must become more intense as depth and distance from the sun increase. Vinogradova's most important results are as follows: Of the 1031 species living on the ocean floor at depths greater than 2,000 meters, 84% are confined to a single ocean. Endemism (amount of strictly local distribution) increases with depth, although genera and families are quite widespread. Deep endemic species are often ancient ones that seem to be preserved in the stable environments of the ocean depths. Other information suggests that speciation and selection for extreme niche stereotypy is a characteristic of fauna of the deep benthos. These results suggest how species diversity can be great in stable environments. First, ancient species and group are retained. Second, small irregularities in topography act as isolating barriers and promote speciation. Third, in stable environments selection is mainly density dependent. Niche divergence of species is encouraged by k selection.

Importance of short-term environmental heterogeneity. The second reason for geographical difference in diversity is that the number of species that can adapt to any particular set of conditions may be inversely related to temporal variation characteristic of the region. The number of species occupying an area is a direct function of the number of species that *could* adapt to it. We know that genetic adaptation to temporal variation involves what may be a fairly complicated set of genetic and biochemical processes. Adaptation is behavioral in some species (Chapter 2). Many species of animals, particularly birds, solve their thermal problems by migrating when conditions become too rigorous or food is scarce.

Adaptation may also include life cycle and physiological and biochemical components. For example, small species with short generation times often have diapause eggs or other life cycle stages that are used to "sit out" unfavorable (dry, wet, hot, or cold) periods. In contrast, large warm-blooded animals generally are able to acclimate physiologically to fluctuations in important environmental variables. Physiologists believe that, other things being equal, a larger animal is less subject to temperature stress than a smaller one. Small mammals are relatively subject to heat and cold stress, and they acclimate poorly in a physiological sense. Methods of adaptation mentioned here and in Chapters 1 and 2 must have evolved with low probability. Therefore likelihood of successful evolution must have been inversely related to degree of environmental heterogeneity.

Persistence in a fluctuating environment might involve genetic tracking (Chapter 2). Tracking species are faced with constantly high levels of genetic death; compensation for high mortality rates is not always possible. Thus we might expect few species to ultimately survive in temperate areas with yearly temperature fluctuation. Adaptation to violently fluctuating

conditions ought to be quite rare since the adaptations mentioned above are probably the result of involved sequences of genetic modification. Therefore, the temporal diversity of temperate zone habitats may have acted to keep animal species diversity at a low level.

The environmental variation faced by tropical species is of a quite different character and magnitude. Temperature fluctuations are relatively minor and occur on a daily period. There are wet and dry seasons in tropic areas, which may pose problems for animals and plants. However, it is probably safe to suggest that tropical environmental variation is less severe than that characteristic of temperate areas.

Janzen has analyzed why the tropics have more endemic species than the temperate regions. (Endemic species are those that occupy a limited and often characteristic area.) Species must be finely tuned to maximize their fitness in a stable environment. Such "tuning," as we know, may involve both loss of genetic variability and loss of ability for further adaptation. Janzen believes that fine adaptation to limited sets of environmental conditions prevents passage of organisms through areas characterized by more extreme values of physical parameters. He has characterized these barriers as being similar to mountains, which are actual physical barriers to population interchange. A mountain range is usually crossed at low places or passes. Mountain pass height in Janzen's terms refers not to actual altitudinal distances but to temperature, rainfall, or other barriers a species would have to surmount during a migratory experience. It is logical that finely adapted tropical species would "see" a difference in some environmental variable as a more impassable barrier than would a species whose genetic and adaptational history includes exposure to more environmental heterogeneity. From the mountain pass analogy we can conclude that the tropics have many species because they allow specialization of species. Once "mountain" passage occurs (establishment of a new colony) a colony may diverge to form a new species, rather than interbreed with the parent colony.

Environmental heterogeneity includes more than

just physical factors. Of at least as much importance to animals and the chance that any particular animal species will persist in an area is the character of vegetation. Vegetation is important not only as a source of shelter and microclimate; it also provides food for herbivores. In this regard there are two major factors that help to explain why species diversities are higher in the tropics than in temperate regions. First, there is always a source of vegetation available to herbivores. This source of vegetation is varied in character; while some trees flower and produce seeds at one time of year, others set seed at different times. During the course of a tropical year, many plant species bloom and provide food for herbivores and seed eaters at many different times of year.

Some of these plants serve as only part of the diet of a variety of long-lived species, with the result that many species are able to persist that otherwise might be forced to leave an area during some season of the year. In addition, many species of plant have their own characteristic fauna. Thus it seems that stability leads to diversity and diversity may well lead to further diversification at other trophic levels. In particular, where a stable climate and highly adapted species exist, the chance that more species will be found ought to be increased immensely. Existing species may be highly adapted to special, specific environments. This observation implies that although colonization of new habitats might be difficult for a species, reunification of the incipient species once colonization has occurred is even more improbable. Without possibility for genetic interchange, colonies are free to diverge genetically either in a random fashion or as a result of differential selection.

Benthos community structure. However genetic divergence might occur, the factors contributing to species diversity may be summarized in Sanders' stability-time hypothesis. Sanders reduces all factors to two: biological accommodation and physical control. According to this thesis biologically accommodated communities are those in which physical variables are relatively invariant and nonrigorous. Such communities occur (among other places) in the deep sea where only slight wave wash and temperature change occur. In such areas the problems an organism, species, or population must cope with are primarily biological. Physically controlled communities are those for which the environment is harsh, unpredictable, and variable. For such communities physical conditions are rigorous enough to preclude any intense interactions among species. The species that exist in such physically controlled communities must use most of their energy to combat the physical environment. Nonetheless, it is incorrect to suggest that there are no species interactions in such physically dominated communities. The rocky intertidal zone of the

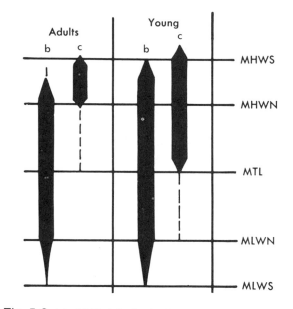

Fig. 5-6. Intertidal height distributions of young and adults of *Balanus balanoides (b)* and *Chthamalus stellatus (c)* at Millport, Scotland. *MHWS,* mean high water level, spring tide; *MHWN,* mean high water level, neap tide; *MTL,* midtide level; *MLWN,* mean low water level, neap tide; *MLWS,* mean low water level, spring tide. (After Connell.)

seashore is one of the most physically rigorous areas associated with the ocean. However, as already discussed, there are a number of predator-prey and competitive interactions in these intertidal areas.

Connell has studied in detail the competitive interaction between two species of barnacle that live on the coast of Scotland. Some of his results are shown in Fig. 5-6. Although the higher intertidal form, *Chthamalus stellatus,* settles and can live quite well low in the intertidal, it is prevented from doing so by *Balanus balanoides,* which characteristically lives in the lower areas. The reverse is not true. *Balanus* is prevented from living in the high intertidal by adverse physical conditions. We can guess that competition between these two species would be far more intense if both could occupy the same environment. They cannot, however, and their separation by physical factors serves as an index to define the many interrelated communities that actually exist in the intertidal zone. Communities are *primarily defined* by the physical variables characteristic of ecosystems. An ecosystem is an interacting unit consisting of all members of a living community plus the values of the physical environment that act on and interact with the living organisms. Such interactions proceed whether a community is physically controlled or biologically accommodated. They will be the subject of much of the latter part of this chapter.

Thorson notes several causes of mortality in biologically accommodated communities. The first of these is related to reproduction. The number of individuals of any species living on the sea bottom is small because nutrients, which must sift down from above, are remarkably scarce. These consist primarily of the remains of planktonic plants and animals and of fishes. Energy reaching the benthos in the form of detritus must have passed through many steps with attendent losses. Because the energy supply is poor, the number of individuals per species must be small. Because of low density it is difficult for individuals to find each other and reproduce. As a result, many of these species are hermaphroditic.

Once zygotes are produced, they must mature to adulthood. The larvae of many bottom species pass through planktonic stages and fall prey to larger plankton, fish, and even whales. At some life cycle stage the remainder of planktonic larval populations begins to search the bottom for suitable substrate on which to settle and spend their adult lives. Often such substrates cannot be found. Bottoms may be too sandy or too silty or perhaps too heavily covered by adults of the species' own kind.

The larva's problems have only started once it finds a suitable place to settle. Thorson calculates that a tremendously high percentage of these larvae are eaten once they become attached to or buried in the soft, stable ocean bottom. Many of the organisms living in the stable bottom community are filter feeders which trap anything coming within reach of their feeding currents or the feeding nets which they construct. Other bottom organisms are detritus feeders. These animals move over or through the bottom mud and sand and vacuum up anything that happens to be in their paths. They "eat" the mud and sand but expell all but organic material. This organic material often includes their own and other young. It is easy to imagine the environment of these creatures as a never ending series of traps, nets, and snapping jaws. Only a few survive to reproductive age. This, then, is the biologically accommodated community: an assemblage of organisms whose primary environmental and adaptive problems are getting enough to eat while avoiding being eaten. No wonder the density of such communities is low.

Fig. 5-7 shows data collected by Sanders from a series of shallow and deep water dredgings. Notice that animal density decreased with depth. Fig. 5-5 indicates that species diversity increases with depth. Why should the number of species increase at the same time that total number of animals decreases? Increases in diversity are related to substrate complexity and diversity. Although the bottom mud that is capable of supporting life is only a few inches thick, its composition varies from the flocculent surface layer to the more

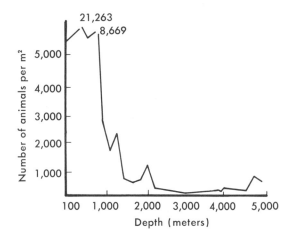

Fig. 5-7. Relation between depth of water and density of the benthic fauna (animal numbers/m²). (Data from Sanders, H. L., R. R. Hessler, and G. R. Hampson. 1965.)

compacted subsurface (both of which are brown and full of organic material) to the gray compacted layer of silt.

Of the species that make up 95% of the total fauna of the shallow ocean floor in Buzzard's Bay nine are detritus feeders, one is a suspension feeder, and one is a carnivore. It is easy to decide why there are so few carnivores in this assemblage. For a predator, net energy gain per food item is approximately equal to gross energy per item less energy expended in search and pursuit activities. In the low density, stable, deep water bottom communities, predators probably spend so much time and energy finding prey that only a few of them can persist.

There are no green photosynthesizing plants on the floor of the deep ocean. Therefore, all animals must depend for their energy source on the green plants that exist at the sea's surface, 1 to 2 miles above. This, of course, means that these animals must rely on dead algae and zooplankton that slowly sink from the sea's surface or that pass through a number of mid-water communities, losing energy with each passage. This is obviously a rather scanty source of nutrient, and we

might expect the species to be adapted to use it maximally. Among bottom animals, 75% feed by grubbing through or over the bottom, 17% are filter feeders taking their nutrition from suspended dead and live animals and plant material, and only 8% are carnivores. Why are deposit feeders so dominant in this community? Suspended food particles may only be present periodically, the result being periodic starvation for filter feeders. Bottom deposit feeders have a far more predictable food supply. Since they can crawl through or over the bottom, they can search out pockets of food and are independent of food periodicity, depending only on the total amount of food that falls on an area of bottom during the year.

The first nine most abundant species of the deep ocean system use their habitat in the ways shown in Table 5-1. One of the two most abundant forms of deposit feeders feeds just at the surface of the sediment where it collects and filters nutrient material from the mud surface and the settling detritus. The second most numerous species feeds by crawling through lower layers of the sediment, ingesting everything in its path, and then digesting only organic material. The third most common species feeds in yet a different way. The reasons why the deep water community is diverse but of low density are as follows: First, the habitat is exceedingly stable allowing persistence of phylogenetically ancient species. Second, the food supply is sparse (and therefore will support few animals) and must be competed for. Because the bottom mud is relatively heterogeneous in character, competition has apparently led to niche divergence and speciation, thereby enhancing original species diversity.

Island biogeography. The study of island biogeography has much to offer a discussion of species diversity and community complexity. Specifically, people interested in island communities have recently begun to explore the problems of island colonization. Imagine a land mass rich in species and from which individuals or seeds can emigrate. Each species has a probability of migration related directly to either its birth rate or population size or both. These emigrants

TABLE 5-1. Niche characterization of the nine most abundant species of a marine bottom community

SPECIES	PERCENT OF FAUNA	FEEDING AND NICHE
Nucula prorina	13.98	Lives just below sediment surface; selective deposit feeder
Nephthys	28.95	Lower gray sediment zone; nonselective deposit feeder (other members of genus are carnivores)
Ninoe nigripes	1.78	Sediment surface; sedentary nonselective deposit feeder (other members of genus are carnivores)
Cylichna oryza	Insignificant	Surface
Callocardia morrhuana	14.97	Surface; suspension feeder
Hutchinsoniella macracantha	Insignificant	Surface; flocculent sediment; nonselective deposit feeder
Lumbtinereus tenuis		Sedentary (like *Ninoe*); other members of genus are carnivores
Turbonilla sp.		Ought to be parasitic; may be a surface deposit feeder
Spio fillicornis		Surface selective deposit feeder

*Data from Sanders, H. L. 1960. Benthic studies in Buzzards Bay. IV: The structure of the soft bottom community. Limmol. Oceanog. **5:**138-153.

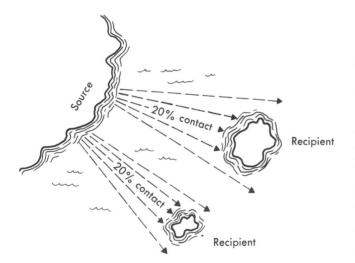

Fig. 5-8. Model of island colonization. Immigrants come from source area and arrive at recipient islands in proportions determined by distance of recipient from source and by size or receiving area of the recipient. Once an immigrant contacts the recipient, its chance of remaining depends on island size, carrying capacity, and niche diversity. Larger islands are expected to hold more species.

are then supposed to float at random in the medium (unsuitable habitat). Eventually these emigrants reach recipient islands and colonize them. Probability of colonization is directly proportional to the perimeter of the recipient island and inversely proportional to its distance from the source. Colonization probability is directly proportional to island perimeter, because larger islands have more beach onto which migrants can land. Two factors contribute to the inverse relationship between distance and immigration probability. First, as shown in Fig. 5-8, recipients near the source will subtend larger arcs of migration than ones farther away. Thus their probability of contact with migrants is greater. Second, migration must involve a per-unit distance chance of death. Since the chance of an emigrant successfully navigating each unit distance is independent of its chance for navigating any other, probability of successful migration is:

$$P = \frac{x}{\pi}(I - d_i)$$
$$i = I$$

d is death rate per unit of migration distance and P is the product of $(l − d)$ for x distance units. These relationships are shown schematically in Fig. 5-8. Once contact of a disseminule has been made, the success of colonization depends on the recipient island's ability to support growth, survival, and reproduction of the colonizing species. Large islands close to the source are expected to have more species than smaller ones or islands further from their source. To see what all this means to species diversity in a community let us expand the model (as in Fig. 5-9).

Imagine a single source and several small recipients versus one large recipient. Initial colonization will be of more species in the case of many recipient islands. This is true because the perimeter of the small islands

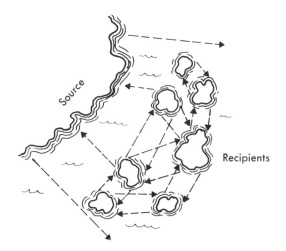

Fig. 5-9. Archipelago model of island colonization. Interchange of species occurs among several islands which are likely to be quantitatively different in flora, relief, etc., and may have different species complexes. If islands are close together, migration will occur at minimal cost and will result in changing interspecific interactions. The more islands, the larger the potential species pool. Increases in the species pool would lead through competition to continuously changing island species compositions. "Islands" may be terrestrial habitat patches.

(taken all together) is greater per unit surface than that of the large recipient. Thus if we assume equal surface areas, the small islands have larger perimeters to intercept disseminules. Each of the small islands is likely to receive and nurture a different set of colonizing species both by chance and because they are slightly different ecologically. As more islands become included in the set, total "beach head" area increases; as island density increases, interisland distance must decrease, with attendant lowering of death rate. Now allow random migration among islands.

Finally make the "islands" patches of habitat in some larger area (forest, cornfield, etc.). Species diversity may be considered as being the result of a patchy environment and of constant migration among patches. Thus we expect most assemblages to consist of two types of species: (1) resident specialists that are highly localized to particular patch types and (2) migratory fauna or flora of generalists, which occur sporadically, perhaps only for a generation or two.

If the model is biologically reasonable, it contains many potential new insights about species diversity, community structure, and stability. If there are many species constantly moving from habitat patch to habitat patch, then competitive relationships are in continual flux. The species composition of a single patch will continually change. If all these species use approximately the same resource, then species Q might displace species A or B or D or X, but leave the others facing a new competitor with which they must share resources. When only two species compete on a long-term basis, we expect them to diverge on their resource. Both could become relatively specialized consumers. However, in this case competitive relationships may not be long standing. We might expect all species so involved to have plastic behavioral niches and generally broad feeding niches. The implications of such dynamic diversity for community stability are obscure.

Data supporting the above discussion were collected by Karr on the avifauna of several habitat types which differ in stability. Species diversity, numbers of

TABLE 5-2. Relations between numbers of bird species and species diversity in tropical and temperate habitats*

| | ILLINOIS | | | |
	GRASSLAND	EARLY SHRUB	LATE SHRUB	BOTTOMLAND FOREST
Total resident species	11	37	46	44
Irregular species †	4	17	13	10
Species diversity ‡	1.54	2.74	3.27	3.31

| | PANAMA | | | |
	GRASSLAND	EARLY SHRUB	LATE SHRUB	MATURE FOREST
Total resident species	25.5	102	109	141
Irregular species†	9	38	41	70
Species diversity‡	1.83	3.46	3.73	3.65

*Data from Karr, 1972.
†Seen infrequently, rare, or waif.
‡$H_s = -\sum_i P_i \ln p_i$, where P_i = proportion of species i; e^{H_s} = number of equally common species.

species, and number of irregular species per habitat type are presented in Table 5-2. Although number of species per habitat increases with environmental stability in the tropics, species diversity (which measures number of *equally common* species) decreases. At the same time proportion of irregular species (which are only occasional visitors to the stable area) increases greatly. These last may be the migrants of the archipelagic model of island colonization.

Although the purely intuitive pictorial model presented suggests why such migrant species should exist in a patchy habitat, it offers no explanation of the relationship between such species and community composition and environmental stability. Neither does it suggest any selective reason for a species existing as a series of subpopulations such as the island model implies.

One relationship between stable habitat and population structure comes from the theory of stochastic (random) processes. It can be shown that when species' birth and death rates are low, as is probably true in stable environments, the species has a better chance of persistence as a complex of interrelated subpopulations

than as a single large group. Chance of persistence is a direct analog of fitness; therefore, such models can be used quite logically to predict ecological and evolutionary strategies.

A second, independent train of thought asks what would be the optimal spatial structure and interpopulation migration rate of a species if that species' environment were patchy and if patches fluctuated in carrying capacity. The originator of the model, Gadgil, analyzed fluctuation in cases where all patches were in phase, out of phase, etc. He found that most forms of fluctuation can be countered best by some interpopulation migration. In particular, when fluctuations in one patch are completely unpredictive of the state of other patches, tendency to migrate should be very sensitive to approach of population size to carrying capacity, but migration rates should be low. When failure of one patch implies increased carrying capacity in others, interpatch migration should be extensive. These two cases may offer some explanation of Karr's intriguing data (Table 5-2).

Simberloff and Wilson have data which also support the archipelago model of tropical versus temperate species diversity. After fumigating some islands in south Florida the investigators observed recolonization events. All islands except the one farthest from the source of immigrants passed through a phase of very high species diversity. Eventually there were lower, competition-moderated species diversities. According to the archipelago model tropical habitats are in constant flux and are expected to have higher than equilibrium numbers of species, which is precisely what Simberloff and Wilson found.

The above helps to explain species diversities in tropical habitats. It also suggests one possible reason for lower species diversities in temperate zones. Temperate forests and fields have fewer species of trees, shrubs, and grasses than do similar tropical areas. In addition, foliage profile diversities (number of foliage layers) are lower in temperate zones. This means that on a per-unit area basis a bird in a temperate climate will encounter fewer units of habitat subdivision than will a tropical species. Habitat islands are further apart in the temperate zone and are often bigger. Trees may be present in large stands dominated by one or two species. Finally, the environment is harsher in the temperate zone. This combination of factors implies high death rate during migration, great distance between islands, and low extinction rates, since islands are large and might be able to support most immigrant species. With low extinction and immigration rate, we expect low species diversity. Karr's data support this expectation.

Much information on field populations strongly suggests that communities in physically harsh and unpredictable habitats are composed largely of waif species; that is, physically dominated communities consist of perpetual, largely unsuccessful colonists. Biologically accommodated communities, on the other hand, consist largely of species that are more niche specialized. According to the island model discussed above, even these are expected to move back and forth among sites and, therefore, may increase the apparent diversity of an area. In fact, quite often rather different communities occur in close apposition separated only

by a transition zone. This intermediate band is called an *ecotone*. Ecotones are often exceedingly rich in species, not only because there are many species specific to these areas, but also because of their transitional character. Most species are waifs. For example, of the 62 aquatic species that have been found in a south Florida estuary only four are estuarine residents. The rest move in and out with changes in water salinity, or they attempt colonization only to die. Still others use the estuary as a necessary habitat for only part of their life cycle. This is particularly true of many commercially important oceanic fish, which spend their juvenile lives in estuaries and salt marshes. Obviously, these waif fauna communities are tremendously diverse.

Succession

In contrast to the relatively stable community characteristic of the deep sea floor, the structure of shallow water assemblages of plankton is quite changeable. Planktonic communities seem to be dominated by the physical factors of their environment. The structure of such assemblages changes rather dramatically with the seasons. In early spring there may be a large number of species; but as the water warms, one species quickly gains dominance, in some cases poisoning other species with its metabolic wastes. Eventually these dominant species become their own executioners, succumbing to self-caused changes in the ecosystem and preparing the way for other species.

The undiscerning observer might suggest that this cycle repeats exactly on an annual basis. However, this is most likely not true. Although changes from year to year may be subtle, they almost certainly take place in all communities. Such orderly and nonrepeating passage of a community through stages is called succession. Most of you have probably noticed this process. A walk across country is likely to take you across an old, previously cultivated field teeming with annual grasses and wild flowers instead of crop plants. As you near the edge of the field you might find the annual species replaced by perennial species (plants that live more than one year and are characteristic of more stable habitats). These species are representative of a later stage of succession than the annual species, and they occur at the edge of the field or in older fields. As fields age they become covered with these secondary species. At the very edge of the field or later in succession grasses tend to be replaced by shrubs. In some parts of the country pine trees begin to fill the fields. In other areas oak trees, or other tree species occupy this successional stage. These trees will eventually form a canopy over the former field. In doing so they will provide sufficient shade for propagation of conifers or hard woods, such as oak, hickory, sweetgum, and perhaps magnolia in the Southeast. Some of these species are viewed as "climax" forms—those that are often found at the end of a successional sequence of changes. (Perhaps this last statement is a bit misleading. Toward the end of succession, it is more likely that the process simply begins to slow down but that change in community composition continues in an orderly manner.)

Climax flora differ drastically from one area to another. In Florida, for example, some uplands are covered with natural expanses of yellow pine, others with small scrubby turkey oak, while wet hammocks contain huge magnolia trees. More northern areas may be covered by maple, beech, and other hard woods. Oregon's Cascade range is covered with several species of fir near the summits and an entirely different set of tree species at lower elevations and in the Willamette valley. On the western side of the Cascades are Douglas fir (which is not a true climax species), but on the eastern slope near the high desert there are sparse forests of Ponderosa pine.

Of course, these climax forests differ for a reason. The floral composition of each is dictated by local conditions. The western Cascade slope has Douglas fir because it has high rainfall and proper soil nutrients. Nearer the Cascade summit more cold-hardy forms take over, only to be replaced very near the peaks by the wind-tolerant and grotesque lodgepole pine. Ponderosa pine grow under dry conditions. Therefore, this tree replaces other species in the Cascades' eastern rain shadow. Progression into the drier desert brings juniper, sagebrush, and rabbit brush to replace the less xerophytic Ponderosa and its associated shrub species, manzanita. We see that three factors—wind, moisture, and temperature—apparently interact to define the tree species of climax- or area-characteristic flora. Soil type is also important.

Southwestern Oregon's Siskiyou Mountains are characteristically dry and hot in summer. They contain several species of flora that are nearly specific to serpentine soil. This soil type is hostile to many plant forms because of its high content of heavy metals, which act as metabolic poisons for many species. It supports its own flora very well, and the flora of serpentine soil is often limited to this habitat. Fungi and bacteria that are usually effective pathogens are unable to grow in serpentine areas, even though they attack serpentine species successfully wherever they occur on other types of soil.

Once community succession has begun, the impact of physical factors upon members of the community is changed and, in most cases, ameliorated. Ecologists have known for many years that plants produce microenvironments that differ quite dramatically from those of barren areas. For example, the night sky acts as an enormous heat sink. The fact that night skies tend to sap heat from living bodies is probably one reason man and certain other animals seek shelter at night. Vegetation, of course, acts to shield animals under its canopy from nocturnal cooling as well as diurnal heating.

Leafy vegetation serves as a moisture source and trap. When plants respire they release water vapor through leaf pores. Because vegetation provides shelter from heat and winds, this released moisture tends to remain in the vegetation-mediated microenvironment. The presence of vegetation also changes the mineral content and porosity of soil. One such change is obvious. A humus layer is formed by a forest through leaf fall or by an open but weed-covered field. The presence of live vegetation fosters conditions suitable for decay and soil production.

EVOLUTION OF COMMUNITIES

It has been suggested that there is evolution at the community level. M. J. Dunbar maintains that oscillations in species abundance, which can occur in unstable communities, are dangerous to the community because species extinction might lead to further instability and loss of community integrity. He concludes that Arctic communities which are based on an unstable environment are evolutionarily immature. In his view, community evolution should involve increased life cycle length of component species and lower rates of increase, both of which should reduce oscillations.

To these criteria we can add: (1) elaboration of food nets, because more links must confer greater buffering against perturbations, (2) greater niche subdivision among possible competitors but *not* loss of overlap, because this would imply resource wastage (MacArthur, 1970), and (3) an increased ratio of biomass (structure) to production (energy flow). These, of course, are the trends we see developing in a succession sequence.

Succession of plant species must be concomitant with animal succession. Each set of plant species produces its own unique form of microclimate that is suitable for specific fauna. What are typical characteristics of species of various successional stages? Fig. 5-10 lists a set of successional trends. Some of these trends are important components of community stability and have implications for pest control (Chapter 6). These trends include the increase during succession of community complexity and stability and the increase in species diversity.

Sources of community stability

There are at least two sources of increased community stability. First, the number of species increases as communities mature. Food webs and competitive interaction become more complex and develop more checks and balances against perturbation. This increase in species number probably results from a matrix of stable microclimates where competitive interaction and selection can result in species subdivision and niche specialization. With specialization comes the possibility of race, subspecies, and eventually species formation as members of a species segregate and form isolated ecological types. This happens at all levels of community structure. In addition, as plant species increase in number, more varied microhabitats become available for insects and other small animals.

A second source of community stability comes from the stabilizing action of the plants themselves. Winds decrease dramatically within a forest or with approach to ground level in a grassy field. A third component of community stability, which changes with community maturity, is the net fecundity distribution of the resident species. As communities become more mature and resident species more k-selected, death rates and number of young born per breeding interval decrease and length of reproductive life increases. The lower rates of increase probably result in greater stability. If all species have low rates of increase, then none will be able to expand its population fast enough in response to changed environmental conditions to out-compete other species and cause loss of community stability.

Animals living in late successional stages tend to be larger than those from less mature communities. Of equal importance is the fact that large, long-lived residents of mature communities must be well buffered against small environmental fluctuations. If each species can maintain a stable population size and all interact closely, then the community must be stable.

ENERGY FLOW

Communities differ not only in microclimate and animal species, but also in the number of trophic levels (feeding levels) that they contain. Many tropic levels are possible. Among the trophic levels that might be included in any community are: (1) autotrophic plants,

On rock early vegetation is moss, lichens.

On sand early vegetation is grassy.

Harsh environment, poorly defined microhabitats—generalist species; plants are annuals, low, grassy.

High plant production, low biomass—organisms are vegetative reproducers or have high birth rates (and usually high death rates), "waifs"—highly migratory.

Plant cover becomes perennial, bushy; microclimates formed; species diversity of plants increases as does within-habitat diversity; humus layer begins to build; species less "waif" like; biological accommodation increases as environment stabilizes, increasing the importance of spatial habitat differences.

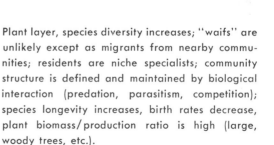

Plant layer, species diversity increases; "waifs" are unlikely except as migrants from nearby communities; residents are niche specialists; community structure is defined and maintained by biological interaction (predation, parasitism, competition); species longevity increases, birth rates decrease, plant biomass/production ratio is high (large, woody trees, etc.).

Fig. 5-10. Trends in succession. (Photos from Kucera, C. L.: The challenge of ecology. 1973. St. Louis, The C. V. Mosby Co.)

particularly those that form organic material from carbon dioxide, solar energy, water, and nutrient minerals (these are also known as producers), (2) herbivores, and (3) one or more layers of carnivores. Herbivores and carnivores are also known as heterotrophs, or "other feeders" because, unlike autotrophic green plants, they must consume other organisms to procure energy. Also included within every community is an exceedingly important group of *decomposers*. One major function of these fungi and bacteria is recycling of nutrient material from plant and animal remains back into the soil or water in a condition in which they can again be assimilated by green plants.

Two factors are important in ecosystem metabolism. These are (1) efficiency of energy transfer and (2) nutrient cycling. Efficiencies of energy transfer are of interest because they can help to define why communities have varying numbers of trophic levels.

Energetic efficiencies of communities begin to differ from each other at the first energy transfer step when solar energy is converted to chemical energy by producers. The percentage of solar energy used varies tremendously among natural systems. Odum estimated that 1.2% of incident energy is converted into plant biomass by aquatic weeds at Silver Springs, Florida. In contrast, the algae of Cedar Bog Lake, Minnesota, have an ecological efficiency (net growth/energy consumed) of only 0.1%. Marine estuaries have producer level ecological efficiencies of about 1.4%. Herbivores generally have ecological efficiencies of about 10%, but these vary widely, ranging from 8.7% (Lake Mendota, Wisconsin) to as high as 30% for marine zooplankton.

There are two components of ecological efficiency. The first component is the percentage of available food that is actually assimilated by consumers. This percentage may be quite low for terrestrial chewing herbivores or those that eat whole leaves, etc., because much plant material is woody or made of cellulose. Both the lignin of wood and the cellulose of plant cell walls are generally undigestible. Thus, much plant biomass in unavailable because of its woody character.

Herbivores like aphids and leaf hoppers which suck sap from their host plants are usually far more efficient than chewing forms, since most of their food can be digested and assimilated. Aquatic herbivores that consume algae get a relatively large return for their investment and are quite efficient.

A second important component of ecological efficiency is loss of energy due to respiration. This is realtively low in herbivores but higher in carnivores. Carnivorous animals usually have higher assimilation efficiencies than do herbivores because their food is already packaged in a usuable form. Carnivores have a high rate of respiration because of the energy they must devote to pursuit and because of their high activity level and consequent loss of energy as heat. The study of an old field in Michigan, revealed that plants lost 15% to respiration, herbivorous mice lost 68%, and the top carnivores, weasels, lost 93%. Of the 47.1×10^8 kilocalories per hectare that entered this particular system as sunlight, 130 k cal. per hectare were incorporated in weasel biomass. This community has only three trophic levels because there is not sufficient energy for a fourth level.

In comparison, offshore marine communities have as many as six trophic levels, primarily because of the high efficiencies of energy transfer at the herbivore level and beyond. Production is quite inefficient because a large fraction of incoming radiation is scattered and reflected by the water, but energy that is photosynthesized is almost all used either directly or secondarily as coalesced protein, fats, and sugars. Some marine organisms may be able to absorb small organic molecules through their skin and membrane surfaces.

Energy production in most communities is the function of green plants. However, in streams, lakes, and estuaries conditions are often unfavorable for green plant life. Yet we sometimes find extremely dense populations of herbivores and animals of high trophic level in such communities. With no plant life indigenous to such areas, where does the necessary energy come from?

It is often thought, erroneously, that the only function of decomposers is recycling of nutrients. However, it has recently become obvious that decomposers serve as agents of energy transfer in certain systems which otherwise have little or no incoming energy. They do so by attacking bits of detritus (such as dead leaves and grass stems) that fall or are transported into the water mass. By attacking these sources of energy, the bacteria and fungi of decomposition release carbohydrates, proteins, and lipid molecules to the water mass. The energetic pathways of these are various. Some of this organic material coalesces into droplets as a result of mutual attraction of the organic molecules and repulsion of water. These globules of organic molecules are then used as an energy source for small zooplankters, which under other conditions, would be herbivores grazing on phytoplankton. Other detrital materials coalesce on solid detritus particles such as zooplankton skeletal debris. These larger sites of deposition then serve as sites for further bacterial action and further deposition. They are the eventual energy source for relatively large herbivorous zooplankton. It has been found that in certain deep Midwestern reservoirs, hebivorous fishes consume detrital material directly when ice cover precludes production by phytoplankton. This material is the refuse of the previous summer's photosynthetic activity and production.

Dependence of herbivores (and indeed entire communities) on detritus as an alternative energy source is of great importance for the continuation of zooplankton populations in far northern areas in which light conditions allow plant growth for only a few months each year. A phytoplankton bloom begins when light and temperature conditions become favorable for growth. As the phytoplankton species increase in density, populations of herbivorous and carnivorous zooplankton begin to grow. As these populations grow, excreta and dead animals and plants form a rather dense crop of detritus. This detritus, because it has a relatively slow sinking rate, remains in upper water levels long enough to sustain zooplankton populations and the community even after phytoplankton production has ceased. Russian workers have found that up to two-thirds of community metabolism can be based on detrital substances derived from a phytoplankton bloom. In addition, many small herbivorous and carnivorous members of acquatic communities feed directly on aggregations of bacteria. In fact, the bacterial aggregations often appear to be significant sources of vitamins for the consumers.

Nitrogen cycle

Whether decomposers serve as sources of community energy or not, they always fill the primary function of nutrient cycling. All nutrients, such as carbon, nitrogen, phosphorus, and sulfur, cycle to some extent through ecosystems, passing from producers through a succession of herbivores and carnivores and finally, via decomposers, back to the inogranic part of the ecosystem where they again become converted to forms in which reuse by plants is possible. Such cycles are often short circuited, but typical examples of several cycles are discussed below.

The pathway that nitrogen follows through a community is shown diagrammatically in Fig. 5-11. According to this diagram, nitrogen can be assimilated only in its fully oxidized form, NO_3. However, no generality is always true; some marine phytoplankton are able to use this nutrient as ammonium, the form in which it is often excreted. Once nitrogen is assimilated, it becomes incorporated into proteins and nucleic acids and is passed from autotrophs to heterotrophs. As passage occurs, excretion and death followed by decomposition release the elements to the medium. These are released first as short-chain polypeptides, then as single amino acids, and finally as NH_4. There follow four bacterial oxidation steps culminating in the eventual reappearance of nitrogen as nitrate. These bacterial oxidations often take as long as 4 months for completion. In the ocean the oxidizing bacteria are active primarily in the dark. As a result, much nutrient material which might otherwise be released in shallow water where production is possible is

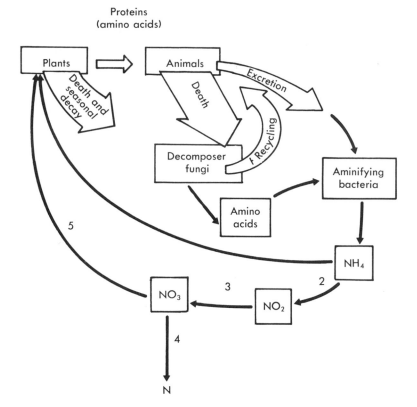

Fig. 5-11. Generalized nitrogen cycle. Proportions of nitrogen that travel between steps are indicated by arrow width. These proportions approximate averages over known systems. The cycle contains numerous bacterially mediated steps, 2, 3, 4, and 5. Much of the plant and animal material passes due to death and excretion to decomposers. Some is immediately recycled in the ocean and terrestrial systems, passing on to herbivores which consume the decomposers. The rest is broken down to amino acids and eventually either enters plant tissue or is lost through leaching, runoff, or conversion to gaseous nitrogen.

only completely cycled after it has sunk well below the photic zone.

Some of the nitrogen that sinks to great depths eventually returns to production via upwelling. A fraction is inevitably lost from the cycle. Such loss also occurs in terrestrial systems but through soil leaching and runoff. Land, if covered with a thick layer of vegetation, slowly absorbs water. In such land, nutrients such as inorganic nitrogen tend to remain near the soil's surface (usually within reach of plant roots).

If, however, ground cover is removed, water sinks rapidly into soil carrying inorganic nutrients with it to relatively great depths. In addition, if the land slopes appreciably, surface runoff occurs. This acts to transport nutrients downhill and eventually into streams and lakes. Improper cultivation causes land to continually lose nutrients. As a consequence of runoff and of compensatory fertilizing, nutrients can accumulate in excess in bodies of water, where they stimulate algal growth. The effects of such stimulation are referred to as eutrophication. Since accumulation of plant material is a natural course of events for fresh water bodies, perhaps it would be more accurate to refer to algal stimulation as too rapid eutrophication. Whatever name we assign to the process, its effects are often serious because of a side effect of overly rapid growth of plant populations. This side effect is rapid respiration, which can materially reduce the oxygen content of lakes and ponds. Such reduction of oxygen is often severe enough to cause almost complete mortality of fish populations, especially during winter when ice cover prevents replenishment of the oxygen.

Nitrogen from lake basins is often returned in small quantities to the land by birds that consume aquatic species. Some is also converted to elemental nitrogen, which escapes to the atmosphere. This nitrogen is returned to the community via either lightning (which causes the formation of NO_2) or as a result of the nitrogen-fixing metabolic activities of algae and certain bacteria which occur in symbiotic association with legumes and some other plants.

The nitrogen cycle is the most complex of all nutrient pathways. It is seemingly open to disruption at several points. Some ecologists have suggested that artificial fertilizers and pesticides might some day be found to poison one of the important bacterial components of the cycle. Regardless of whether or not such a drastic consequence of man's activity is possible, we are obviously causing some rather severe imbalances in community structure with regard to nitrogen. Leached land and eutrophication are two correlated effects of our activity. In addition, we have the ecologically amazing habit of disposing of our sewage into rivers and other convenient large bodies of water. This seems ridiculous because two better alternatives are available.

Land is many times more capable of disposing of organic wastes than is the best river-sewer. Experts estimate that a maximum of 100 acres per thousand population is required. This amount of land could process all municipal and industrial sewage. Other studies have indicated that in some areas 4 acres per 1,000 population would be sufficient. Green belts, wildlife sanctuaries, fire breaks, and most agricultural lands might profitably be converted to fertilization by direct application of municipal waste—at least that from which harmful bacteria have been removed. Use of wastes as fertilizer could thus serve two important purposes, recycling of nutrients with avoidance of water pollution and fertilization of farm lands. One inherent danger of such disposal methods is the possible contamination of ground water. Other esthetic problems might need solving as well. They city of Milwaukee has found it profitable to dry treated sewage and market it as high grade organic fertilizer. The Chinese have been fertilizing with ''night soil'' for centuries.

As previously mentioned, nitrogen can be used by phytoplanktonic organisms as ammonia. In estuaries such short-circuiting of the nitrogen cycle is carried to extremes. Plant debris such as the leaves, stems, twigs, and flowers of terrestrial plants form the base of production in such systems. The leaves, for example, are initially broken down by amphipods which graze upon them and physically tear them into tiny pieces. Other primary breakdown is by bacteria and fungi, which remove carbohydrates, lipids, and nitrogen compounds. The nitrogen compounds are excreted by the bacteria as short-chain polypeptides and amino acids. In this form they are available for further use by bacteria. The bacteria that utilize the suspended nitrogenous substances use leaf debris as a substrate. Thus nitrogen and other nutrients are concentrated on leaf debris as a result of bacterial activity. This concentrated nutrient is then grazed by zooplankters and some fishes. The leaf particles, having been left relatively intact by their consumers, again act as nutrient deposition and concentration sites. Much of the nitrogenous material thus concentrated is probably cycled many times before it is broken down to more elemental forms such as ammonia.

Other biogeochemical cycles such as those for phosphorus, sulfur, and carbon have been described. Rather than discuss these I would like to make one fairly obvious but critical point. *All of the cycles and energy transfer rates are interwoven and interdependent.* Each nutrient (and raw energy) is a potential limiting factor of an ecosystem. One of these must be limiting at all times unless space is truly limiting. The mussel *Mytilus edulis* concentrates nutrients for its dependent community. These nutrients are used directly by consumers rather than by plants. Loss of the *Mytilus* would be serious to the intertidal organisms for which it concentrates vital nutrients.

Ecologists have been aware for some time that, at least in closed communities, disruption of any cycle could result in complete loss of community function. Complete loss of function is rare, but instances of partial loss are obvious everywhere. Nutrient cycles are often broken in cultivated land because much of the crop is exported and residues are sometimes burned. Farming, therefore, becomes progressively more costly because of the necessity of treating fields with fertilizers to restore their nutrient material. Land that has had its vegetation removed for one reason or another is often a victim of surface runoff, erosion, and leaching. These physical consequences of denudation sharply decrease the land's potential productivity.

In most cases communities are not entirely separate. Neighboring communities interact to form larger intertacting structures. For example, the marine benthic community depends for nutrients on the plankton directly above the bottom, which in turn depends on plankton nearer the surface. Nutrients of phytoplankton come from the land and to a greater extent from the ocean deeps. Some of the energy transformed by the surface community is exported to the land and perhaps back again eventually to the sea. Because there seem to be many interactions among sets of communities, with sufficient time and inclination one could interrelate all communities found on the earth's surface.

One type of community interaction is through transfer of materials and energy. We have seen several examples of this. Communities also share space or exclude one another from space. Open fields displace woodland systems, but are soon replaced by natural assemblages, should man fail to contribute sufficient energy for their maintenance. As long as species pass freely from one community to another, we must consider the assemblages to be functionally linked. Anadromous fishes link the sea with mountain streams that are hundreds of miles away.

Beginning with the individual and ending with global patterns of community relationships, ecology is the study of strategic and evolutionary interactions. Individual animals interact with their environment, form populations, and in so doing interact with members of their own species. Members of diverse species interact to form dynamic communities, and these in turn act to alter their environment and consequently provide physical and chemical bases for interactions among themselves and for their own succession.

BIBLIOGRAPHY

Connell, J. H. 1961. The influence of interspecific competition on the distribution of the barnacle *Chthamalus stellatus.* Ecology **42**:710-723.

Dunbar, M. J. 1960. The evolution of stability in marine environments; natural selection at the level of the ecosystem. Amer. Naturalist **94**:192.

Finenko, Z. Z., and V. E. Zaika. 1970. Particulate organic matter and its role in the productivity of the sea. In J. H. Steele. Marine food chains. University of California Press, Berkeley.

Gadgil, M. 1971. Dispersal: Population consequences and evolution. Ecology **52**:253-262.

Golley, F. B. 1968. Secondary productivity in terrestrial communities. Amer. Zool. **8**:53-59.

Janzen, D. H. 1967. Why mountain passes are higher in the tropics. Amer. Naturalist **101**:233-250.

Khailov, K. M., and Z. Z. Finenko. 1970. Organic macromolecular compounds dissolved in sea water and their inclusion in food chains. In J. H. Steele, editor. Marine food chains. University of California Press, Berkeley.

Karr, J. R. 1971. Structure of avian communities in selected Panama and Illinois habitats. Ecol. Monog. **41**(3):207-233.

MacArthur, R. H., and J. W. MacArthur. 1961. On bird species diversity. Ecology **42**:594-598.

MacArthur, R. H., and E. O. Wilson. 1967. The theory of island biogeography. Princeton University Press, Princeton, N. J.

Margalef, R. 1961. Communication of structure in planktonic populations. Limnol. Oceanog. **6**:124-128.

Odom, H. T. 1957. Trophic structure and productivity of Silver Springs, Florida. Ecol. Monog. **27**:57-112.

Paine, R. T. 1966. Food web complexity and species diversity. Amer. Naturalist **100**:65-75.

Pianka, E. R. 1966. Latitudinal gradients in species diversity: a review of concepts. Amer. Naturalist **100**:33-46.

Sanders, H. H. 1968. Marine benthic diversity: a comparative study. Amer. Naturalist **102**:243-282.

Sanders, H. L. 1960. Benthic studies in Buzzards Bay. IV. The structure of the soft bottom community. Limnol. Oceanog. **5**:138-153.

Sanders, H. L., R. R. Hessler, and G. R. Hampson. 1965. An introduction to the study of deep-sea benthic faunal assemblages along the Gayhead-Bermuda transect. Deep-Sea Res. **12**:845-867.

Simberloff, D. S., and E. O. Wilson. 1969. Experimental zoogeography of islands. The colonization of empty islands. Ecology 50:278-296.

Simberloff, D. S., and E. O. Wilson. 1970. Experimental zoogeography of islands. A two-year record of colonization. Ecology **51**:934-937.

Teal, J. M. 1962. Energy flow in the salt marsh ecosystem of Georgia. Ecology **43**:614-624.

Thorson, G. 1966. Some factors influencing the recruitment and establishment of marine benthic communities. Netherlands J. Sea Res. **3**:267-293.

Vinogradova, N. G. 1959. The zoogeographical distribution of the deep water bottom fauna in the abyssal zone of the ocean. Deep-Sea Res. **5**:205-208.

Zenkevich, L. A., and J. A. Birstein. 1956. Studies of the deep water fauna and related problems. Deep-Sea Res. **4**:54-64.

6 APPLICATIONS OF THEORY

Humanity, as a vast species population, is well beyond the point of being able to get away with wasting resources. Globally, the tillable land mass is becoming inadequate for our sustenance. As recently as 1970 demise of the world's oceans as a result of pollution was predicted at scientific meetings. Red tide (destructive blooms of *Gymnodinium brevi* and related dynoflagellate algae) is occurring in polluted ocean areas where it has not been reported before. Current estimates are that we are overfishing almost all commercially import finfish populations. Among these the Atlantic salmon is in jeopardy because of Danish fishing efforts. We wage a continual battle against insect pests, which lower the quantity, quality, and market value of our crops. Several of these pest insects are developing resistance or immunity to general pesticides; the pesticides themselves are destroying the esthetic value of our environment and at the same time are lowering its ability to support our population.

It is not the purpose of this chapter to detail the vast store of ecological maladjustments that are caused directly by man's careless, naive use of his environment. Rather, I will attempt to show alternatives to our present policy. Some of these are already in use on a small to medium scale. Others are only theoretical ideas whose application has yet to be tested in the real world. The theories we have discussed in previous chapters can be applied to fisheries and wildlife managements, mariculture, and control of agricultural and forest pests. All of the techniques and management strategies invoked are based on ecological and, to a certain ex-

tent, economic theory. The material on community structure is of overriding importance to pest management. Therefore, we will treat each major area in turn beginning with strategies of game and fish populations management.

STRATEGIES OF GAME AND FISH POPULATION MANAGEMENT

Not too many years ago, hunting was allowed at all seasons of the year, and there was no limit to the amount of game an individual could take. Seemingly limitless herds of buffalo and vast flocks of passenger pigeons were driven to near extinction and extinction. As other game animals, notably large ducks like the pintail and canvasback, also began to decline, state and national governments began establishing and enforcing regulations designed to slow or arrest the decline. There were and are three basic regulations. The first prohibits hunting during nesting or parental nurturing seasons, the second either prohibits or discourages killing of females, and the third sets bag limits designed to harvest excess animals without detriment to the population. However, even with fairly enlightened regulations, some species (particularly ducks and mountain sheep) continued to decline. Investigation has shown that the major reason for such decline is deterioration of breeding grounds in the case of ducks and of winter grazing territory in the case of bighorn sheep.

Solving these problems requires money rather than

strategic analysis. Ducks Unlimited is an international organization that buys and protects endangered Canadian duck breeding lands and restores previously drained grounds. The Bureau of Land Management is making some efforts to remove cattle from bighorn sheep winter range. Habitat restoration and management are also used to "restore" pheasant (an introduced species) to several areas and are practiced at great expense on several quail hunting plantations in northwestern Florida.

Such means of increasing crops of game animals is not the real subject of our discussion. It is of more interest to investigate how populations are or might be managed to maximize yearly yield. The scheme is conceptually simple, but it requires a detailed knowledge of the reproductive capabilities of the species or population being studied. The procedure can be followed with reference to Fig. 6-1, which shows a series of relationships between present population size and the size expected a year later, given the initial size. This model is based directly on the simple discrete generation model of population growth in a limiting environment (see Chapter 2).

Points on the graph falling above the 45 degree line indicate net increase in population size (from time t

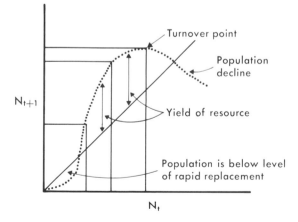

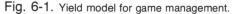

Fig. 6-1. Yield model for game management.

when N_t parent animals are in the population to time $t + 1$ when the population consists of both adults and their young of the year). The curve of N_{t+1} versus N_t begins at a low point when, for example, breeding might be curtailed by low numbers of adults and consequent inability to find mates, or when winter kill of adults is excessive because of small population size. (This is particularly true for birds and animals that depend on each other for heat or warning of predators.) Thereafter, yields increase with increased adult population size until the carrying capacity of the environment is exceeded. At this turnover point yields begin to decrease.

To interpret such a model we need to note that the diagonal line passing equidistant between the axes of the figure denotes exact replacement of the population ($r = 0$). The population size resulting from some initial population size after 1 year is derived graphically by drawing a vertical line up from the horizontal axis at some population size and then drawing a horizontal line from the point at which the vertical intersects the population growth curve to the vertical axis. The intercept of this line on the vertical axis is the expected population size. The potential sustained yield of the population is the segment of vertical line between the break-even line and the population growth curve. Therefore, maximal yield occurs at the point where the growth curve turns over and begins to decline.

The use of such models requires that management personnel have detailed knowledge of (1) the ability of a range to support a population, (2) the species' intrinsic rate of increase, and (3) what might be expected to happen to reproduction over a set of years with the normal amount of variation in weather and habitat conditions. Given this sort of knowledge, populations can be effectively managed so as to maximize yield to hunters and to minimize wastage and suffering of game due to winter starvation. Prevention of winter starvation and overgrazing also protects range quality.

Simple models like this one are adequate for scientific management of game populations when growth is determinate and when gross reproductive potential is not a function of age. However, their inadequacy is striking when dealing with the management of fisheries stocks and yields. In this case another important variable is added to the management equation and an old variable is greatly complicated. The new variable is fish size and growth rate; the old one is reproduction. Instead of fish being taken at some mature size as are deer, they can be harvested at virtually any size, from fingerlings to large (but rare) adults. Harvest of young fish precludes their reproduction and endangers the entire population. Therefore, the first strategy of fisheries management must be either to catch only fish that have had some chance to reproduce or to catch only a limited number of fish of reproductive size. There must be a breeding pool large enough to guarantee population replacement. Salmon can be harvested to fairly low population levels without harm to recruitment levels. The former strategy must apply to pelagic fish stocks or to those whose young are not as space and food limited as the salmon.

The second strategy takes account of the observation that most pelagic fishes are limited by food supply. Therefore, it behooves the fishing strategist to harvest fish at or near the age when energy ceases to be devoted largely to growth. Generally, fish growth is logistic; young, small fish grow rapidly until they reach some pivotal size. Beyond this critical size growth rate decreases until growth is exceedingly slow in very large members of a species. These large, unproductive fish still use energy but with little net return to the fisherman and at some cost to the growth rates of younger, potentially more productive animals. Therefore, fish should be harvested before lowered growth rates are noticeable. Of course, this part of the management scheme cannot interfere with allowing fish to breed enough times to make population replacement likely.

The final consideration in our yield model (the entire model is shown in Fig. 6-2) is natural fish mortality. Obviously, no prudent, hungry predator can afford to wait and watch all of its prey die or become otherwise unavailable. Neither should it harvest its prey

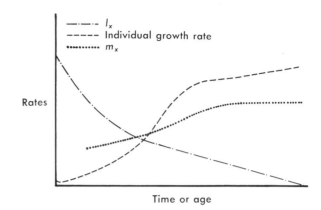

Fig. 6-2. Simple yield model of fisheries management.

before the biomass of an age class (of which eventual yield is a direct funtion) reaches a peak. Peak biomass, considering both natural mortality of fishes and individual growth rates, is at the point where survivorship and size curves cross. Fish populations for which this peak occurs later than onset of reproductive maturity are harvestable on a sustained yield basis. Others, like salmon, must be harvested with escapement, allowing enough fish to reach adulthood to guarantee sufficient recruitment to replace the population.

Management of fish populations is more complicated than the above discussion indicates. What if a population were harvested after the first or second reproductive age class and harvested so heavily that essentially no later age classes existed? Think back to the idea of many reproductive age classes being necessary in an environment that varies in suitability for reproduction. Removal of this stabilizing property of population structure is almost bound to result in violent fluctuations in population size—a complication no operation closely managed for maximum sustained yield can tolerate.

The size of fish harvested can be controlled by enforcement of laws setting a minimum net mesh size such that small fish are allowed to escape. Unfortunately, international regulation of net mesh size is difficult to introduce and more difficult to enfore. Our fisheries are being overfished. Fleets concentrate their efforts on particular areas until they reach a minimum acceptable level of production. Then they move on. The overfished areas sometimes take as long as 20 years to recover their productivity. Presently the Norwegians are seeking to extend their fishing territorial limits from their coasts, because they claim that fishermen of other countries are overfishing the codfish that they can most conveniently reach. While there is no doubt that the codfish population is declining, the question of who is overfishing remains to be answered. Scientists have determined that the cod's reproductive rates have decreased along with an apparent increase in mortality and decrease in average size. Some suggest that this indicates overfishing, while others claim that natural changes in the environment are at fault. The Atlantic salmon, an important sport fish in this country and the British Isles, is another overfished species, and currently is in danger of extinction. Adult fish congregate in a particular, small area of the North Atlantic where they can be fished in an efficient and concentrated manner. Too few escape to breed later in North American and British streams, and the population is declining at an alarming rate. An agreement has recently been reached with the Danes, who fish salmon at sea, to curtail pelagic netting in the hope that the fish may be saved.

Similarly, sailfish, marlin, and bluefin tuna are being overfished in some waters by Japanese fishermen who employ long lines to effect capture. These are 1- to 4-mile cables equipped with many baited hooks upon which fish become impaled. Serious depletions of these predominantly sport fish have been noted. One investigator points to a great decrease in 4-year-old sailfish in Floridian and Bahamian waters. Such loss of fish of reproductive age could damage the ability of these populations to maintain their numbers. The catch of fish taken by such methods cannot be easily regulated in terms of harvest of optimal sized fishes. About the best that can be hoped for is establishment of national and international catch limits or protection by maritime countries of their continental shelf populations.

No single resource has been more poorly used than whales. Early in the history of whaling methods used in the capture and processing of these mammals were inefficient enough that the stock of whales was not seriously threatened. However, with the advent of modern fast equipment, populations began to be seriously decimated. When whaling nations realized that their resource was facing extinction, catch limits were instituted. But the resource was one shared by all the whaling nations, and these competing interest groups were unable to reach agreement as to the maximum sustained yield of the fishery (despite practically universal agreement among biologists about the necessary strategy). As a result, the blue whale effectively disappeared as an economical resource in 1960. Following the demise of these largest whales, fishing effort was switched to fin whales. But again the yearly catch limits suggested by experts on whale demography were grossly inflated by international agreement among whale fishermen. The fin whales effectively disappeared within 4 years. Sei whales, to which effort was then turned, were exterminated even more rapidly. The tragedy of this chronicle is that, had the biologists' recommendations been followed, a substantial sustained yield would have been possible.

The above example suggests that the human factor must be considered when setting regulations and determining fishing strategies. Many nations were involved in competition for this particular resource. Each was striving to maximize its own net gain. By the time the blue whales had been decimated many of these nations had tremendous investments in whaling ships and processing equipment. They felt that they could not afford to limit their catch. Of course, they had a valid economic point, viewed in a narrow perspective. Considerations pertaining to the economics of resource utilization are treated in the excellent volume edited by Dorfmann and Dorfmann (see Bibliography).

Clearly, sustained yield management of fish populations will be a losing battle if our population continues to grow and require more protein. In terms of overuse of the sea and aquatic production (some of which is second in yield per unit area only to intensely cultivated agricultural land) we have two "outs". The first of these is general in scope. Wherever and whenever protein can be procured from cultivatable plants, such sources should be used because of production efficiency. Every step up the trophic ladder is only 10 to 30% efficient. Therefore, direct use of plants is indicated on energetic grounds.

Another logical step is to cultivate herbivorous fish. But where and how? Among aquatic-marine ecosystems, estuaries are natively richest in nutrients and energy. Their source of energy is decayed plant material deposited by abundant mangroves, other trees, and by various marsh grasses. Particles of this decaying plant material are covered with bacteria which recruit nutrient materials from the surrounding water. These bacteria are an important source of food. "Where" seems to be at least partially answered. But what about *exactly* where? Will anywhere in an estuary do, or might we take advantage of known physiological characteristics of fish and shellfish? Obviously, most of the ways that the system's yield can be enhanced should be exploited. We find that among fishes and shellfish, mullet, shrimp, and blue crab are all tolerant of high temperatures. In fact, the growth rates of all these species are enhanced by increased temperatures. We are currently faced with the problem of what to do with tremendous quantities of hot water produced by coastal electric power generating facilities. To date much of the thermal load has been released into estuarine creeks and salt marshes with regrettable effects. However, management personnel of many of these plants are beginning to answer the questions I have just posed. Several plants in Florida are raising shrimp in cooling ponds. When these animals are raised in warmer than normal water, they grow rapidly and sometimes reach incredible sizes. Blue crabs and mullet can be raised in the same way. All three species provide high-quality protein. Power plants along the Gulf Coast are experimenting with other fish such as pompano, sea trout, drum, and croaker. Success with most of these seems certain, particularly since the future holds the possibility of breeding most commercial species for fast growth.

Many fish whose life cycles include—and perhaps require—an offshore, pelagic stage may prove refractory to mariculture. However, juveniles of no fewer than twenty species of fish are obligate inhabitants of lower East Coast salt marsh creeks and estuaries, and juvenile life cycle stages are among the most expensive to a species in terms of mortality. Perhaps if juveniles of these species were raised under optimal conditions or were otherwise encouraged during their estuarine existence, the yield of coastal fisheries could be enhanced. Stocking salmon and steelhead rainbow trout has met with great success in Washington, Oregon, California, and the states surrounding Lake Michigan. The hatchery principle might profitably be expanded to include establishment of natural hatcheries. The state of Maine has been stocking lobsters since 1960. No one knows whether this effort is aiding a declining fishery.

The stocking of hatchery-reared fish, particularly trout, is commonplace in heavily fished waters. At first, fisheries biologists noticed a disturbing phenomenon: introduced fish, whether of the same or different species as natives, usually suffered high death rates. Plantings of trout fry in lakes or streams with resident breeding populations usually resulted in insignificant return to anglers. So did planting of legal trout in the fall, after fishing season. Even planting on a strictly put-and-take basis results in only about 35% return of planted fish. Most returns in put-and-take stocking are during the first 2 weeks after a plant is made. Almost none of the introduced fish survive into their second summer.

Some of the death which apparently follows stocking is caused by shock. Fish that have been reared in hatchery ponds are ill-equipped to face the rushing waters of natural streams. Hatchery fish can be taken easily from quiet backwaters but not from riffles or the main current of a stream. In contrast, fish that were exposed to stream conditions before release into a wild population showed considerably better survival.

Most of the mortality of hatchery reared trout cannot be explained by differential initial survival. Evidence has long suggested that there is a significant component of competition between released fish and natives. For several years the idea of competition was discounted by fishery workers because trout are able to survive long periods of starvation. Competition with wild trout for food should have little effect on newcomers.

Miller challenged the contention that death of introduced fish is not due to competition. He compared mortality and weight gain in hatchery trout when placed in streams with significant populations of wild fish and when placed in streams in which wild fish had

been removed. He found that on the average mortality of the introduced fish was 55% in trout placed in competition with wild populations, but only 12.9% among fish not forced to compete with natives. In similar experiments Miller noted that during the first few weeks after planting, hatchery fish appeared lethargic. They could be easily caught by hand. Also, hatchery trout placed in competition with native fish had considerably higher blood lactic acid levels than did either the natives or hatchery trout placed in similar sections of stream from which all residents had been removed. Lactic acid level of the stocked fish declined after 14 days. Within the 14 days following planting hatchery trout, death averaged 72% of their total death rate. In his discussion Miller relates these results to the

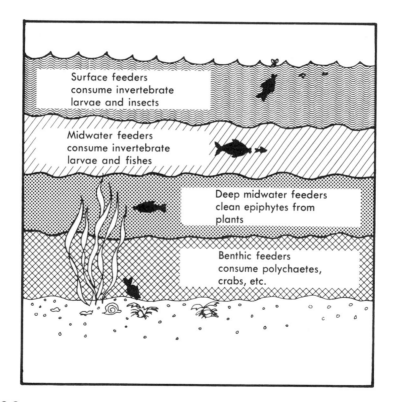

Fig. 6-3. Feeding zones in typical estuary illustrating one way species subdivide their habitat.

fact that trout are territorial. He concludes that much of the death of hatchery trout is the result of competition for territories with already established fish and that death is also caused by starvation, perhaps acidosis, and fatigue. These results sharply limit the situations in which fish stocking is of value. Such stocking would be beneficial in streams in which the natural trout population density is quite low and in lakes and streams that have been reclaimed by destruction of native populations.

One question that should be answered is: Should fish farms contain "monocultures" with each pond containing only one species (as do some tremendously successful Louisiana catfish farms) or is there an advantage to mixed species farming of fishes? At least a partial answer is provided by Javan fish culturalists. They find that ponds containing nine species of carp (carefully chosen to avoid competitive interactions among species) produce far more protein than similar ponds with less fish diversity. We have already investigated the principle involved here. Remember that *Drosophila simulans* and *D. melanogaster* reach larger total population size when cultured together than when raised as single species cultures. They use different portions of the medium. The same principle can apply to fishes in a mariculture or natural situation.

Carr has defined the presence of five different vertical feeding zones in a body of water not more than 6 feet deep (see Fig. 6-3). Each of these zones is characteristically used and occupied by a different complex of juvenile fishes. In addition, three classes of fish found on the grass flat can coexist with minimal interaction. These are eaters of algae epiphytic on marsh grass, detritus feeders, and cleaners, which pick parasites from other fishes. At least two of these might profitably be raised together. In addition, since no fish population is likely to consume its entire food supply (detritus), addition of other species that consume herbivorous zooplankton might be indicated. Even better, different species of juvenile fish consume food particles of different sizes. Because detritus particles are of range of size, a variety of species that consume different kinds of debris might be grown together with good results. Such speculation is not idle. The natural marshland fish community divides its resources in all of these ways but, in addition, suffers high mortality rates. Mariculture operations are and will be designed to maximize yield by maximizing individual growth rates and minimizing death. Such operations will require the cooperation of ecologists, geneticists, physiologists, and engineers for their eventual success. An appreciation for the value of mariculture can be gained from the following figures. Lake Erie produces a maximum of 7 kg per hectare per year. Fertilized catfish farms produce an annual yield of 169 kg per hectare, West Javanese carp ponds yield 500 metric tons per hectare, and an offshore yellowtail fishery produces 280 metric tons for the same surface area. Taiwanese milkfish can be grown at the rate of 2,000 metric tons (2,000,000 kg) per hectare per year, and oysters cultured in Japan yield 58,000 kg per hectare per year. Notice that the most productive artificial systems are those that cultivate herbivorous fishes.

Other than cultivation of estuaries, the possibilities of raising nutrients from the ocean are rather limited. Only farming of the continental shelf, generally no more than 100 miles wide, makes real sense. Perhaps the most well-known efforts to raise marine organisms is oyster farming. This really only amounts to increasing the spatial carrying capacity of the environment by providing suitable banks or beds of substrate. The same principle has been applied to game fish that require shelter from predators and hunting grounds for prey. Natural rock outcrops are the usual source of this limiting factor, but recently game and fish departments and sportsmen's groups have built artificial reefs. These are large piles of paving material, cement-filled rubber tires, or old automobile bodies. These reefs greatly increase concentrations of sport and game fish.

Inland streams can be greatly improved by enhancement of habitat diversity. Low partial dams or groins built at intervals along an unproductive stream cause pools to be carved from the bottom by increased water velocity. These serve as sheltering holding water for trout. Fine gravel lifted from pools is then deposited to form riffles which support dense populations of aquatic insects. Trout then thrive.

On the other side of this particular coin, stream canalization, the practice of straightening stream channels as a means of flood control, is an unequivocal ecologic disaster. Surveys have shown that because of increased silt load and loss of pools and riffles, canalized streams lose about 90% of their productivity of fish. Such streams typically have such swiftly flowing current that they are unable to drop silt as do normal rivers. As a result, some downstream body of still water gets these materials, often to its detriment. The effects of canalization are not limited to the stream and its fish. The very act of canalization ruins streamside forests and fertile bottomlands, removing wildlife habitat. Recently conservation groups have successfully resisted some canalization plans.

Whereas canalization unduly increases rate of stream flow, the building of reservoirs or trapping of rivers for large-scale irrigation can also be harmful. Rivers that are heavily tapped for irrigation water tend to have such reduced flow that large fish kills occur periodically because of anoxia. Normal flow results in constant aeration as water flows over riffles and small rapids. Reduced flow reduces aeration potential and also allows temperatures to rise. The effects of reservoirs on aquatic communities are mixed. Many excellent trout streams have disappeared forever behind dams, but tailwater (below dam) fish production and production in the reservoirs themselves are often excellent.

Dams do tremendous damage to populations of anadromous fish like salmon. Adult fish are unable to get upstream except with the aid of fish ladders, and young fish returning to the ocean are trapped behind hydroelectric dams and swept through the power generating turbines. But the most gruesome effect of Columbia River dams on salmon was only realized a few years ago. Water emitting from these dams is saturated with nitrogen so that salmon tissues become super-saturated with dissolved nitrogen. When the fish reach normal water the nitrogen forms bubbles within their blood, giving them the ''bends.'' Most of these fish die.

Hyodroelectric dams and any industrial facilities, including nuclear power generating stations that use river water for cooling, increase downstream temperatures. This produces a variety of ecological effects including lowered oxygen content and increased eutrophication. Low oxygen content can prevent passage of fish to upstream breeding sites.

At least one study demonstrates well the effect of reservoir building on estuarine ecology. This is particularly interesting because it indicates how seemingly remote systems are really interdependent. The oyster is resident in marine estuaries and seemingly occurs in large numbers only in those estuaries that drain a land mass and thus receive periodic fresh water. The reason for this involves complex ecosystem interactions. Investigators have determined that the oyster parasites and predators, *Dermocystidium marinum,* boring sponges, the carnivorous snail *Thais haemostoma,* and the drills *Urosalpinx* and *Eupleura* are all driven from estuaries or killed by low salinities arising from periodic freshwater runoff. When this runoff is curtailed, as by reservoirs, oyster populations suffer greatly from predation and parasitism. In addition, oysters may not be able to breed when salinities exceed 40 parts per thousand. (Normal sea water is 35 ppt, but evaporation from an estuary can raise salinities considerably.) Thus at least one important marine species is heavily dependent on normal flow of fresh water and is damaged by the short range economic expedient of stream damming.

Shrimp are also dependent on fresh water flow into their estuarine nursery areas. This commercially important group (the Gulf Coast catch was worth $60 million in 1962) arrives in the nursery grounds at the same time as spring and fall fresh water runoff peaks. The juveniles require an external source of vitamin B_{12} which is high in runoff waters. Population densities have been found to be highly correlated with average annual rainfall. Decreases in catches off Texas following a 1950s drought period were not followed by population regeneration in some populations. At least one author attributes this fact to incidence of dam and reservoir building on contributory freshwater streams. The blue crab is also dependent on fresh water and its populations decline perceptibly during periods of drought.

River flow and usage are large, complicated problems of economic strategy. What is needed is a balance between short-term immediate gains and losses and long-term effects of environment-altering action. For example, damming a tidal river is expected to contribute x dollars to local economy purely as a result of the labor force required to build and maintain the dam. In addition, the dam might provide hydroelectric power of monetary value. It might also serve a flood control purpose and provide a constant supply of water for irrigation. All of these positive effects can be given a per-year economic value. Added to the equation or cost benefit study are several negative terms. The dam and reservoirs by their mere presence will destroy a section of river which may have great esthetic value and value to sportsmen. In addition, fertile farm or timber land will be inundated and families will be displaced. Spawning runs of anadromous fish will be disrupted and their populations will be endangered. Estuaries are damaged, as already mentioned. Many of these terms in the equation are easily thought of on a short-term basis. Long-term effects might include siltation of the reservoir and closure of the estuary's passage to open sea. Although the equation is not particularly difficult to write (at least in simple terms, before depreciation rates are considered), it is rather difficult to attach dollar values. We might hope that our solution of it will proceed at a faster rate than the evolutionary rate at which natural populations solve resource utilization and energy partitioning equations! It is clear that any environment-altering action should receive careful forethought and consideration of ensuing losses and benefits before implementation. Similar accounting must apply to land fill projects, building of heat generating power plants, and marsh draining.

Legislation now requires that corporations and governmental agencies file environmental impact statements before they proceed with some potentially degrading project. Theoretically, these should be powerful protectors of the environment if properly researched and reviewed. Unfortunately, people can be found to perform biased research and provide glowing cost-benefit ratios.

FOREST MANAGEMENT

Methods of timber harvest also lend themselves to analysis of strategies. There are several ways to harvest timber, ranging from the massive clearing of virgin forests (which occurred in this country during the last century) to today's modern intensive forest management. The pertinent questions here are: At what age should a forest be cut? How large should cutover blocks of land be? Should clearcutting or selective methods of harvest be followed? Naive answers to these questions are as follows: The trees should be harvested when they have reached their growth plateau and before they can be seriously damaged by pests. They should be harvested in blocks as large as possible because massive cutting tends to minimize setting up time (road building, etc.). Clearcutting should be used because it is far more efficient than selective cutting.

The naive answer to the first question is absolutely correct. Trees, like fish, should be harvested after their growth years have passed and before insect and plant parasites and pathogens can make serious inroads into the economic value of the forest. Senescent trees are more likely to be victims of such pathogens than are young, actively growing specimens.

Answer number two is manifestly incorrect. The damage that large-scale clearcutting has done and is still doing is awe inspiring. In the Siskiyou Mountains denudation of mountain slopes, erosion, loss of nutrients, and stream siltation and damage are results of massive timber removal. Tropical rain forests suffer similar fates when indiscriminantly harvested. In addition to these effects of large-scale clearcutting, large cutover areas regrow as stands of even-aged trees which are often perfect habitats for insect pests, although they do not provide the necessary habitat diversity or type for their predators, parasites, or competitors. Similarly, game populations are reduced. Thus great long-term loss can result from massive timber cutting. Of far more utility is the practice of cutting small patches of timber on a yearly rotation basis. In the South patches are cut about once every 30 years and patches of various ages are contiguous. Habitat patchiness and diversity are therefore maintained at a high level. Suitable habitat is available for a variety of insects, including pests *and* their enemies. The system, therefore, remains relatively stable, and losses of production to pests are kept at a minimum. This is another instance of the practical importance of habitat diversity. In addition, patchy cutting increases the ability of timberland to support game animals, an aspect of real value to the sportsman and nature observer.

The third question has really been answered already. Clearcutting is extremely unpleasant esthetically, but it is economically sound. When practiced with discretion (on flat or gently sloping land away from streams) and in small patches, it provides necessary timber economically, it is not detrimental to future yield, and it actually provides superior habitat for game.

AGRICULTURAL MANAGEMENT

Can any of the principles outlined above be carried into modern agriculture? The answer, of course, is "yes." In fact, any principle presented thus far has application in agricultural systems, particularly with respect to pest management and yield maximization. The rest of this chapter will deal with problems and principles of economic insect pest control.

For some reason, which we will explore below, most crops are plagued to some extent by herbivorous insects. These animals damage crop plants in various ways, piercing leaves and sucking sap, consuming leaves, boring holes in stems, and consuming seed or fruit. Often such insects are too sparsely distributed or cause too little damage to be of economic importance. Others, however, may operate above the economic threshold, at a density beyond which further density and more damage will outweigh the cost of insect control. It is these common insects, whose pattern of population density fluctuation is characterized by outbreaks, that I wish to consider.

Several pertinent questions need answering. (1) Are "pests" common in natural systems or are they rather specific to cultivated stands? (2) What causes an economic pest? Can any features of insect-ridden areas or communities be consistently noted? (3) How might pest insects be controlled? How might they not be controlled?

In respect to the first question, harmful insect outbreaks very seldom occur in virgin forests or in other undisturbed areas. This is particularly true of communities with naturally high species diversity. Arctic areas, with their sparse fauna, do have outbreaks of lemmings which would certainly be classified as pests if man had much of a vested interest in Arctic vegetation. Further south, in Russian forests and steppes, outbreaks of rodents and other potential pests are less common. Since floral and faunal diversity increases rapidly with distance from the poles and since the increase is accompanied by reduced likelihood of violent population fluctuation, it seems reasonable to conclude provisionally that outbreaks of possible economic importance are related to species diversity. We might expect to find that reduction in diversity, disturbance of an area, or cultivation results in instability of some species populations. Since rapid response to changed conditions can be expected only of species with high intrinsic rates of increase, we expect insect outbreaks. Indeed, such is the case.

As early as 1908 it was realized that: "Man, in planting over a vast extent of country certain plants to the exclusion of others, offers to the insects which live at the expense of these plants conditions highly favorable to their excessive multiplication." (Marchal,

1908.) Weevils more often attack white pines growing in pure culture than those growing with other trees. Numerous other examples demonstrate that cultivation or pure culture leads to organisms assuming pest status because of excessive density. Two reasons for this monoculture phenomenon might be suggested. First, monoculture results in an unbroken expanse of host plants. With no barriers to intrapopulation migration and breeding, insects are able to approach their intrinsic rates of increase. In contrast, life in small pockets of suitable vegetation, as in a diverse mature forest, would imply mating difficulty and possible local insect starvation and subpopulation extinction.

Second, recall the discussion in Chapter 5 of the rocky intertidal community. The potentially dominant, encrusting form, mussels, were kept from dominance by a series of predators, by wave wash and battering, and to a lesser extent by competitors. This natural community was relatively stable. The principle of complex food webs and trophic interrelationships applies equally well to other communities. Natural communities, because of their diversity and complex interrelationships, are stable and contain no overridingly dominant species. In cultivated communities, on the other hand, herbivorous insects have fewer predators and parasites, and they are able to realize much of their intrinsic rates of increase.

Pimentel provides a nice demonstration of the influence of plant species diversity on animal species. He grew *Brassica oleracea* in three plots: (1) mixed species planting (*B. oleracea* with naturally occurring open field plants), (2) single species *(B. oleracea* alone), and (3) sparse, single species. Pimentel then censused the fields at weekly intervals during which he determined species diversities and numbers. Outbreaks of aphids occurred in the single species planting, but in the second year the aphid population was far more subdued. Trouble with pests was less severe in the more diverse plantings.

Pimentel's work and observations, which have been made on natural systems, give us this precept: Insects and other organisms generally become economically important pests when natural systems are either intrinsically simple and unstable or when the structure of communities is simplified (hence made less stable).

Curiously enough, it is safe to say that pests are most often created by man. The ways in which this happens all involve disturbance, some of which is unwitting and some of which is purposeful. Pimentel's work showed us one way in which pests are created. Artificial formation of a simple unstable community will probably be favorable to only a few species. These are likely to be herbivores and are more likely than predators to adapt to specialized conditions because of their generally large birth rates and attendant abilities to evolve rapidly. Released from predators and competitors, such species are free to become economically significant. A dramatic example of pest creation comes from the USSR where the virgin steppe of Kazakhstan has only recently been converted to wheat production. Studies of this system showed three things. (1) Species diversity of the wheat fields was less than half that of nearby virgin grass land. (2) Insect population densities of the wheat fields were nearly double those of the undisturbed areas. (3) Most of the wheat field species were important pests.

Pests are also made by introduction. Organisms are likely to be part of a dynamic balance when in their own communities. Some of these same innocuous species, when carried or allowed to disperse to new territory, can become numerically dominant and, if on crop plants, pests by definition. Many of California's wild grasses are introduced species, as are the ubiquitous and obnoxious starling and English sparrow. So too is the rabbit of Australian fame. In this country the cotton boll weevil once lived in Mexico, but irrigation patterns and planted crops allowed its migration to southern cotton country. This species probably has been the recipient of more insecticides than any other insect. There are about 10,000 pest species; many are pests only because of transplantation and attendant loss of density-governing predators, parasites, and competitors.

Pesticides make pests! This happens in two ways. First, a naturally controlled potential pest species can be released from control through pesticide-caused death of its controlling agent (a predator). Precisely this happened with scale insects of California citrus orchards. The scales were a problem to citrus growers in the nineteenth century until ladybird beetles were introduced. The beetles preyed upon the scales and kept them in check until 1946 when citrus groves were sprayed with DDT. This reduced predator density and scale insects once again became an economic problem. Reestablishment of control took 3 years from the time use of DDT was abandoned. In another study pesticides were shown to reduce the number of species in a community, raise the ratio of herbivores to predaceous forms, and increase the number of insects.

"Upset pests" are another creation of chemical pest control. These insects are normally present in low densities but are released from control by pesticides applied originally to control some other species. A classic "upset" species is the red spider mite, a pest of fruit orchards. This species was first released by tar oil washes which killed its predators and destroyed their habitat. Later releases were caused by DDT, which nearly exterminated most of 40 naturally occurring insect species and left the field open to red spider mites.

Now that we know that pests are made, not born, our next problem is obviously one of control. You may have already concluded that most pesticides are to be avoided if possible. There are a number of reasons for this strategic policy. Among them are: Most insects (particularly ones that evolve rapidly) can build up resistance to lethal chemicals, as shown in Table 6-1 and Fig. 6-4. In Fig. 6-4 we see that the yearly increase in resistance to pesticide is truly remarkable. Among the resistant forms are, obviously, species that are most commonly sprayed, such as houseflies and mosquitoes. Table 6-1 illustrates the phenomenon of cross-resistance. Insects sprayed with one chemical often become resistant not only to it but also to related poisons. There are four groups of pesticides within which cross-resistance is possible. Resistance is the result of natural (or unnatural) selection for rare resistant mutants. Since there are only four groups of insecticidal chemicals, thought suggests that only time is necessary for the production of totally resistant insect

species. (We have already discussed the role of pesticides in pest release and formation of upset pests.)

All available evidence points to the importance of certain pesticides as environmental pollutants. DDT is persistent and worldwide. Oceanic birds and pelagic

TABLE 6-1. Relative resistance to insecticides acquired by citrus red mites exposed to Systox compound (the mites are 266 times more resistant to Systox than would be a normal population)

COMPOUND	INCREASE IN RESISTANCE	RESISTANCE RELATIVE TO SYSTOX
Trithione	15,000	56.4
Iso-Systox	8,570	32.1
Methyl-Systox	8,000	30.1
Schradan	2,000	7.5
EPN	1,400	5.3
Parathione	833	3.1
Phosdrin	350	1.3
Pyrazoxon	300	1.1
Systox	266	1.0
Tetram	111	0.5
Delnav	100	0.4
Malathion	8	0.0+

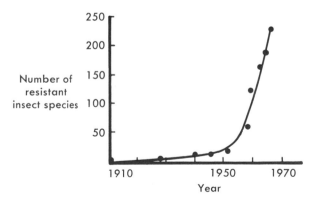

Fig. 6-4. The growth of resistance to insecticides. The number of resistant species refers to the number of different insects and mites with resistance. (After Conway.)

fishes contain it. The chemical has been strongly implicated as the cause of failing populations of pelican, peregrine falcon, and other birds, and of declines in some fish populations because of reproductive failure. While pesticides will probably never be phased out completely, better ways to use them and better alternatives for pest control exist. Among these are:

1. Timing of pesticide use and use of selective pesticides
2. Breeding for host resistance
3. Genetic control of pests via "tricks"
4. Biological control—introduction of disease, predators, competitors, and parasites
5. Cultural control
6. Integrated control
7. Control involving life history studies and optimal time for attack with pesticides

Pesticide timing

Once the benefits of pesticides like DDT were well known, farmers in some areas began using the chemicals to excess. The effects of such policy were often disasterous, as illustrated by the Cañete Valley story (from G. R. Conway, 1971). In 1920 this Peruvian valley was devoted largely to cotton growing. Three pests were economically important by 1940, but these were controlled relatively well with inorganic poisons. In 1949 the growers began to apply DDT when a rather serious loss due to bollworms and aphids was noticed. This practice was successful at first, but then growers determined that if a little pesticide was good, more must be better. In order to make application easier, barriers to dusting, such as trees, were removed (simplifying the community ever further), and insect predators, birds, and parasites of the herbivorous pests were reduced to extremely low levels. The various herbivores began to develop resistance to the pesticides, and control was almost completely lost with frequency of treatments rising to one every 3 days. Finally, the cost of pesticides exceeded yield! More rational programs of control could have avoided the disaster. Such programs typically involve maintaining pest population densities and applying chemical

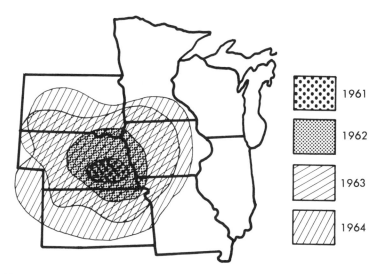

Fig. 6-5. Spread of western corn rootworm resistance to dieldrin and related insecticides. (After Conway.)

poisons to minimize population density at some critical period, for example at flowering time or when fruit is setting.

The densities and population growth rates of some pests are highly dependent on weather conditions. Adequate knowledge of pest biology combined with predictive meteorology allows efficient use of pesticides by spraying only during years of favorable weather. If pests can be hit hard with pesticides just before potential population outbreaks, population growth can be effectively curtailed. Remember that increase in population density depends on initial population size. Timed programs of pesticide application have cut usage by about 50% in Colombia's cotton growing regions. In so doing they have lessened the chance of resistance developing and spreading through the population. Rate of change in gene frequency is directly related to selection intensity.

Pests are pests only because they have escaped control by natural density regulatory agents or because natural agents, such as predators, fail to keep pest levels below economic thresholds. Since general pesticides often kill predators faster than they do herbivorous pests, selective use of pesticides which harm only herbivores is indicated. When pests and their predators have asynchronous life cycles or live for parts of their life cycles in separate areas, chemicals can be applied to harm only the pests.

The use of truly selective pesticides is still not well developed. It should be possible to develop and use chemicals which affect only particular classes of pest organisms (such as sucking aphids or herbivores that ingest plant material). A few such chemicals that have minimal effects on natural control agents are in use, but development of others is bound to be slow since research and development are very expensive.

Pesticides can often be applied in conjunction with pest-specific baits, thereby avoiding the possibility of disturbing natural control mechanisms. One such bait is a member of the opposite sex. Dr. Max Witten of Australia's Commonwealth Scientific and Industrial Research Organization (CSIRO) suggests that female

insects be raised in the laboratory, selected for resistance to some insecticide, and then sterilized. These baits, which have lost none of their attractive powers for the males, would then be liberally loaded with the insecticide and released. Of course, all males that mate with these flying Mata Hari's would be killed, and sex ratios would decline to such low levels that the wild population's ability to reproduce would be seriously hampered.

Breeding for resistance

This method of pest control depends only on the genetic and physiological characteristics of the host organism. Hosts can be developed that contain antibiotic substances damaging to pests, that have thick skins, that are structured to prevent pest penetration, and that are simply resistant to pest attack. Breeding programs often involve crossing desirable, but nonresistant varieties of crop plants or animals with less desirable, but resistant varieties. The resulting genetic line is then selected for its original desirable characteristics, but it is also resistant to the pest. Over 95% of our grain acreage is planted to resistant varieties, but there is no reason for complacency with this means of pest control. Many pests of grain plants are fungi and bacteria. These have shorter generation times and higher rates of increase than their hosts. Therefore, new pathogenic varieties continually arise for which resistance must be developed. Also, the development of resistant varieties often involves a long period of inbreeding, which must reduce the plants' genetic variability and may result in lowered overall fitness unless attention is given to this aspect of the breeding program.

Genetic control of pests

Genetic control is exceedingly interesting and takes advantage of many of the principles discussed earlier. Most means of control discussed here are temporary, but all can be devastating to the pest organism. Perhaps the most temporary genetic "trick" suggested involves introducing genes from southern populations into more northern populations of the same insect species. Alleles from the sourthern variety might be those which determine length of development time, temperature-dependent mating behavior, or some other evolutionarily derived life cycle or developmental trait. Introduction in high frequency in spring, for example, if handled properly, would result in high frequencies of the gene by fall whereupon all homozygous carriers would fail to complete their life cycles during cold fall months. Thus an extraordinary source of mortality might be imposed on the northern population. Reversal of this trick is also theoretically possible. Northern genotypes introduced to sourthern populations might result in rapid development during cool spring weather (perhaps when no economically important host plants are present and populations are at low densities). These would then be at considerable selective advantage over their southern counterparts. Frequency of the northern genotype would rise in the spring, but there might be high mortality if the northern genotypes were adversely affected by hot weather.

A rather more detailed method of genetic control is being tried by CSIRO. Individuals heterozygous for chromosome translocations have low fitness, because meiotic events usually result in partial sterility. A program of blowfly control involving release of five or more different types of fly each carrying a different translocation is the subject of experimentation. Theoretically, the translocation flies released will be homozygotes. Release of all five will be simultaneous and in swamping quantities, but in different areas so that future matings will result in formation of translocation heterozygotes. Under optimal conditions, one can anticipate mortality rates approaching 98% . Following initial mortality, pockets of translocation homozygotes are expected to remain. As these expand, they should come in contact and a new bout of mortality should occur. Such control should remain effective for several years, after which the population might have to be reseeded with translocation homozygotes.

Genetic methods offer intriguing possibilities for control of species which are amenable to laboratory rearing programs, and it is probable that most economically important species will lend themselves to culture.

Control by habitat manipulation

Habitat manipulation has been called "cultural control" by at least one author. It takes many forms, ranging from habitat alteration designed to disrupt a pest's life cycle to manipulation designed to encourage a pest species' natural enemies. Mosquito control as practiced in some areas is an example of life cycle disruption. These insects lay eggs in still ponds, tree bole puddles, and other bodies of water. When marshes and ponds are drained or covered with a film of oil, the mosquitoes' life cycle is disrupted. However, this seems to be ecologically harsh. These same marshes support fish populations and also serve as breeding sites for water birds.

A less disruptive example is control of white pine blister rust. This fungus attacks pines, but requires barberry to complete its life cycle. The National Park Service has pursued a program of barberry removal in order to control the pest.

One reason cotton pests became disastrously important in Peru's Cañete Valley was that farmers practiced ratooning (leaving cotton plants standing for 2 or 3 years). This obviously gave the pest insects a stable environment in which population growth could continue year round. Clearing the fields following each growing season disrupts populations of many pest insects, thereby preventing establishment of large populations. Crop rotation is a varient of this procedure. Most pests are host specific so that alternation of crops can effectively trim pest populations by limiting the time during which they can grow. Growing two crops simultaneously in a field can often provide enough spatial heterogeneity to prevent rapid growth of pest populations, both by reducing reproductive rates and by providing possible refuges for predators and parasites.

Hedgerows are means of cultural control, since they

provide habitat heterogeneity and refuge for natural control agents of herbivorous pests. Species diversity of insects falls off at an exponential rate with distance from a hedgerow. Many of the insects composing this diversity are control agents of herbivorous forms. The hedgerows provide a heterogeneous environment in which many such forms can prosper and from which they can invade the specializing and generally inhospitable field.

Herbivorous insects are generally smaller than their predators and, therefore, they usually have higher intrinsic rates of increase. This fact can often be used in cultural control, particularly with regard to forage crops which are harvested by periodic mowing. Were such crops harvested by mowing an entire field at one time, both herbivores and carnivores would have their homes disturbed. Whereas the herbivore populations would quickly recover, those of carnivores would exhibit signs of damage for much longer. Such mowing practice results in periodic outbursts of pest populations. However, mowing a field in swaths and at relatively long intervals leaves bands of vegetation within which predator densities remain high. The predators can make sallies into mowed strips in search of rebuilding herbivore populations. Such strip mowing can offer constant control of pests.

Biological control

Biological control, which usually involves introduction of predators or parasites to pest populations, has been spectacularly successful in some cases, but one never knows when a control agent will become a worse pest than the one it was introduced to destroy. The mongoose was introduced to Hawaii to control snakes but soon developed a taste for domestic poultry. Rather than cite a large number of case histories of successful biological control, I will discuss only a few, using them to demonstrate control principles.

One of the first spectacularly successful introductions of a control agent was in California. In 1887 the cottony cushion scale, *Icerya purchasi,* was a serious pest of citrus. In conjunction with several parasites, the ladybird beetle, *Rhodalia cardinalis,* was introduced. This beetle quickly reduced scale insects to economically tolerable levels. One of the reasons for its success was its size relative to that of scale insects. A single beetle was able to consume enormous numbers of scale insects. It was successful also because of its mobility. The beetle was able to search out and consume colonies of scale insects nearly as fast as they could recolonize. Both the cottony cushion scale and the ladybird beetle were exotics. While this feature made the scale an economic pest because it lacked natural enemies, at the same time it allowed the beetle's success. With no natural competitors the predators were able to expand rapidly to the detriment of their prey.

One problem is sometimes automatic with introduced predators. A consequence of simple predator-prey systems is extinction of the predators because of eventual insufficient food supply. Introduced predators may have no secondary prey to fall back on once primary prey have been decimated. Therefore, it is often advisable to find alternate prey for introduced predators. Otherwise, the simple predator-prey system must be spatially diverse so that new prey colonies, hence supplies of food for predators, are always available. Success of introduced predators is not guaranteed; such simple systems require fortuitous balances of population dynamics or careful planning and search for suitable control agents.

These principles and others are illustrated in a study of naturally occurring biological control agents done by Huffaker and Kennett. *Tarsonemus pallidus,* the cyclamen mite, is a pest of California strawberries. The system is complex. Besides the cyclamen mite, which attacks growing strawberry leaves causing malformation of leaflets and lowered berry production, there are two other important herbivorous pests. Of these, the two-spotted mite, *Tetranychus telarius,* is active in early spring, weakening plants and increasing their susceptibility to the cyclamen mite. The strawberry aphid, *Capitophorus fragaefolii,* causes little direct damage but may be a vector of a virus disease of strawberries which lowers their fruit-producing ability. Whereas the two-spotted mite produces summer plant conditions which are favorable to cyclamen mites, the virus carried by strawberry aphids lowers the plant's ability to support cyclamen mites. Three predators are important in this system. Of these, the mite, *Typhlodromus cucumeris,* is an effective predator of cyclamen mites. This species can fall prey to two other less effective predators, *T. occidentalis* and *Orius sp.,* which commonly prey on two-spotted mites but may switch to cyclamen mites when the density of these becomes high. In addition to this interaction, the aphid produces honeydew which supports *T. cucumeris* when cyclamen mite densities are low.

Biological control in this system amounts to stocking and close manipulation of *T. cucumeris,* since it is necessary to spray fields in spring for two-spotted mites. Close integration of biological, chemical, and cultural control is necessary for maximum economic benefit.

The cyclamen mite winters in small numbers in strawberry plant crowns. Since the mites require moderate temperatures and relatively high humidities, they do well only within microclimates afforded by unopened or partially opened leaves. Thus populations never become really large even when predation is absent. Reproductive rates of *T. cucumeris* are very similar to those of the mites. In addition, the predators are good at searching for mite concentrations, they have similar seasonal patterns of abundance, and they are generally well adapted to maintain control of their prey. These predators are more motile than their prey and trap prey which move from infestation foci, thereby limiting the prey to refuge microhabitats where their numbers cannot be large.

Manipulation of predators is necessary because they do not naturally gain control of the pests until the second (and economically important) year of infestation. This pattern results in periodic heavy damage by mites, which lasts for part of the second year of infestation. Avoidance of this control lag is the major manipulatory objective. Obviously, control must involve stocking first year fields with predators in order to minimize the lag phase. In tests made by Huffaker and Kennett, stocking did result in early establishment of a predator-prey equilibrium and economic control of pest mites. However, there is a disadvantage to first year stocking. Control of two-spotted mites is made more difficult because use of acaricidal chemical disrupts predator populations as well. An "out" to this is second year spring stocking. This, too, proved to be effective. Predator-stocked plots usually had less than 10% the pest populations of predator-free plots. The authors suggest that stocking of new fields be done by seeding them with material obtained from normal winter pruning of older plots (those with equilibrium proportions of predators and cyclamen mites). In this way both predators and sufficient food for their support can be introduced to maximum advantage.

Biological control is no panacea for pest problems. It often fails eventually. (Recall the Myxoma virus that was introduced in Australia to control rabbits. Within a short time the two species co-evolved to live in harmony.) Failure is not the worst problem with introduced control agents, however; there are several cases in which the "control" species became a pest itself. Some weed control agents may someday develop a taste for valuable species of plants, or a predaceous insect may attack other, natural control agents. Introduced control agents are tested to reduce their chance of causing unexpected problems, but short-term testing takes no account of the possibility of genetic changes which would allow the control agent to switch hosts or prey. Nor is it considered that predators introduced to control a particular species of pest suffer from tremendous intensities of k selection every time the pest population is reduced to low density. Today's helpful control agent might well become tomorrow's plague.

Of the methods of pest control discussed, cultural means of control may be the most desirable. All the other methods are aimed at either extermination or rigid control of a single species. None of them can have effects limited to the target organism. Pesticides are notorious for widespread damage and disruption of community structure. Agents of biological control can themselves become pests. Even genetic means of control can be dangerous because they are designed against only one species. Reduction of one pest can cause outbreaks of its competitors. We must conclude, therefore, that cultural control is preferable, since in its more enlightened forms it is designed to restore community diversity and stability, thereby controlling a complex of possible pests.

In this chapter I have tried to "put it all together." Earlier chapters introduced genetic and ecological principles and, within the scope of any introductory text, indicated the extent to which populations and their component parts and species interact. Through the interaction of species populations communities are formed, only to again interact at a higher level of organization. More than anything else I have tried to emphasize the adaptability of entities at the various levels of ecological organization, and the necessary complexity of structure exhibited by populations and especially by communities. This complexity should be used, not simplified, as we strive to live dependent on ever larger portions of the living world which surrounds us.

BIBLIOGRAPHY

Conway, G. R. 1971. Better methods of pest control. In W. W. Murdoch, editor. Environment, resources, pollution and society. Sinauer Associates Inc., Stanford, Conn.

Dorfmann, R., and N. S. Dorfmann, editors. 1972. Economics of the environment. Selected readings. W. W. Norton & Co., New York.

Huffaker, C. B., and C. E. Kennett. 1956. Experimental studies on predation: Predation and cyclamen-mite populations on strawberries in California. Hilgardia **26**:191-222.

Miller, R. B. 1958. The role of competition in the mortality of hatchery trout. J. Fish. Res. Bd. Canada **15**:27-45.

Pimentel, D. 1961. Species diversity and insect population outbreaks. Ann. Ent. Soc. Amer. **54**:76-86.

Watt, K. E. F. 1973. Principles of environmental science. McGraw-Hill Book Co., New York.

AUTHOR INDEX

SUBJECT INDEX

Trout—cont'd
   rainbow, 45
   steelhead, 91
Turtle, green sea, 73
Typhlodromus cucumeris (mite), 168
Typhlodromus occidentalis, 110-111, 168

U

Urosalpinx (oyster drill), 159

V

Variance
   environmental, 36, 44

Variance—cont'd
   genetic, 26, 36, 44, 46
Viability fitness, 43
Volcanoes, 80
Voles, field, 81, 88
   and predator response, 118-120

W

Wallaby, Australian, 114-115
Warblers, 71, 108
Waste disposal, 149
Weather, 1-5, 74, 78, 108, 153
"Weed" species, 45, 67

Weevils
   cotton boll, 162-166
   pine, 162
Whales, 155
   baleen, 118
White pine blister rust, 166
Wolves, timber, 114-117

Y

Yield
   maximizing, 153
   sustained, 154-155

1440109